Student's Book 4

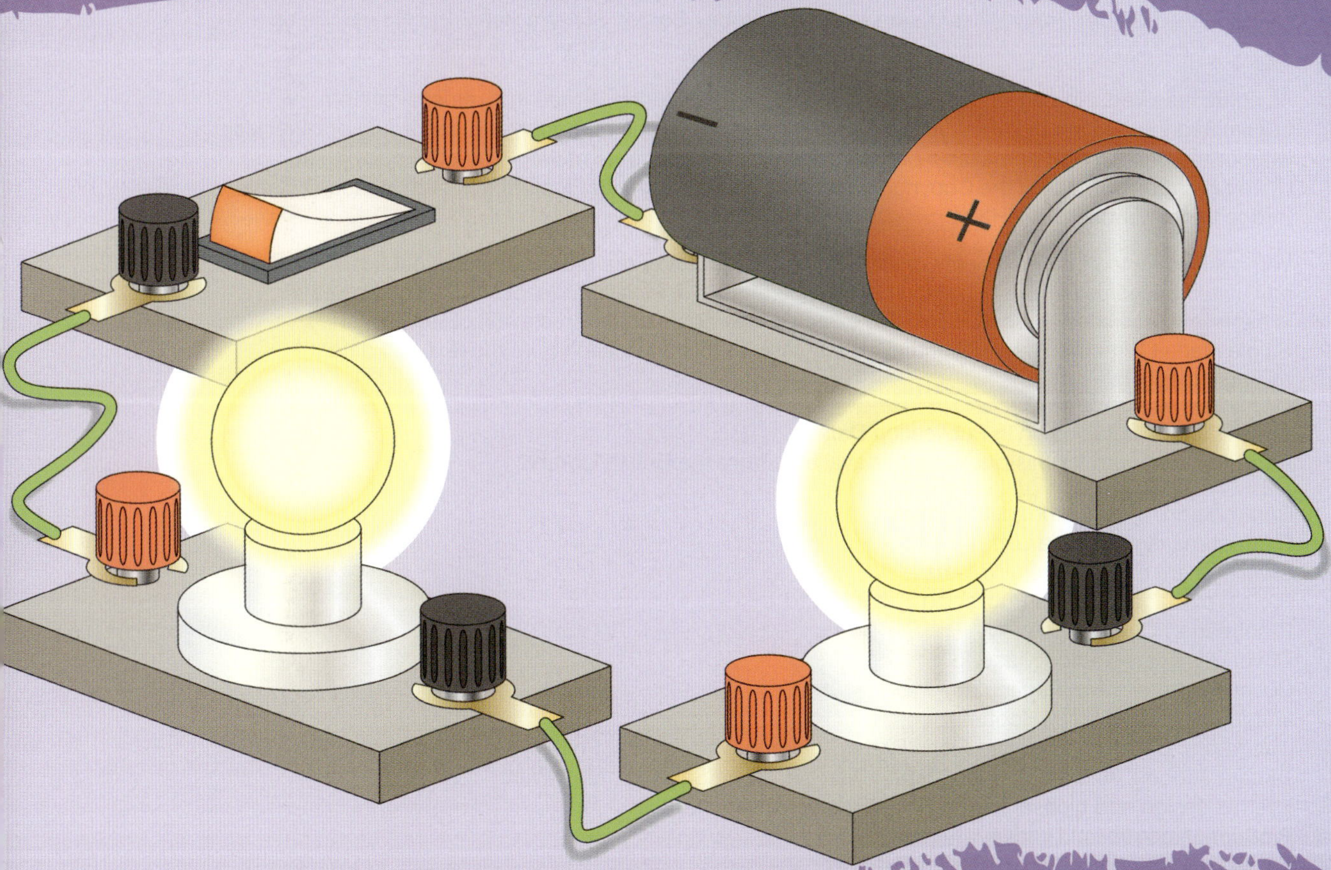

William Collins' dream of knowledge for all began with the publication of his first book in 1819. A self-educated mill worker, he not only enriched millions of lives, but also founded a flourishing publishing house. Today, staying true to this spirit, Collins books are packed with inspiration, innovation and practical expertise. They place you at the centre of a world of possibility and give you exactly what you need to explore it.

Collins. Freedom to teach.

Published by Collins
An imprint of HarperCollins*Publishers* Ltd.
The News Building
1 London Bridge Street
London
SE1 9GF

HarperCollins*Publishers*
Macken House,
39/40 Mayor Street
Upper, Dublin 1, D01
C9W8, Ireland

Browse the complete Collins catalogue at
www.collins.co.uk

© HarperCollins*Publishers* Limited 2021

10 9 8 7

ISBN: 978-0-00-836890-6

Second edition

Contributing authors: Tracy Wiles, Karen Morrison, Tracey Baxter, Sunetra Berry, Pat Dower, Helen Harden, Pauline Hannigan, Anita Loughrey, Emily Miller, Jonathan Miller, Anne Pilling, Pete Robinson.

All rights reserved. No part of this publication may be reproduced, stored in a retrieval system, or transmitted in any form or by any means, electronic, mechanical, photocopying, recording or otherwise, without the prior written permission of the Publisher or a licence permitting restricted copying in the United Kingdom issued by the Copyright Licensing Agency Ltd, 5th Floor, Shackleton House, 4 Battle Bridge Lane, London SE1 2HX.

Without limiting the exclusive rights of any author, contributor or the publisher of this publication, any unauthorised use of this publication to train generative artificial intelligence (AI) technologies is expressly prohibited. HarperCollins also exercise their rights under Article 4(3) of the Digital Single Market Directive 2019/790 and expressly reserve this publication from the text and data mining exception.

British Library Cataloguing in Publication Data
A Catalogue record for this publication is available from the British Library.

Commissioning editor: Joanna Ramsay
Product manager: Letitia Luff
Development editor: Karen Williams
Project manager: 2Hoots Publishing Services Ltd
Proofreader: Caroline Low
Cover designer: Gordon MacGilp
Cover illustrator: Ann Paganuzzi
Image researcher: Emily Hooton
Illustrators: Beehive Illustration (John Batten, Moreno Chiacchiera, Phil Garner, Kevin Hopgood, Tamara Joubert, Andrew Pagram, Simon Rumble, Jorge Santillan, Matt Ward)
Internal design and typesetting: Ken Vail Graphic Design Ltd
Production controller: Lyndsey Rogers
Printed and bound by: Replika Press Pvt. Ltd.

With thanks to the following teachers and schools for reviewing materials in development: Preeti Roychoudhury, Sharmila Majumdar and Sujata Ahuja, Calcutta International School; Hawar International School; Melissa Brobst, International School Budapest; Rafaella Alexandrou, Diana Dajani, Sophia Ashiotou and Adrienne Enotiadou, Pascal Primary School Lefkosia; Niki Tzorzis, Pascal Primary School Lemesos; Vijayalakshmi Chillarige, Manthan International School; Taman Rama Intercultural School.

Registered Cambridge International Schools benefit from high-quality programmes, assessments and a wide range of support so that teachers can effectively deliver Cambridge Primary.

Visit www.cambridgeinternational.org/primary to find out more.

Contents

Topic 1 Life processes and ecosystems — 1

- 1.1 Exercise for health — 2
- 1.2 Medicines — 4
- 1.3 Stopping disease — 6
- 1.4 Vaccines — 8
- 1.5 Habitats for animals — 10
- 1.6 Adapting to different habitats — 12
- 1.7 Food chains — 14
- 1.8 Energy for survival — 16
- Looking back Topic 1 — 18

Topic 2 Humans and other animals — 19

- 2.1 Modelling the human body — 20
- 2.2 The human skeleton — 22
- 2.3 Moving your bones — 24
- 2.4 Functions of the skeleton — 26
- 2.5 Animal skeletons — 28
- Looking back Topic 2 — 30

Topic 3 States of matter — 31

- 3.1 Investigating play slime — 32
- 3.2 Materials, substances and particles — 34
- 3.3 Solids and liquids — 36
- 3.4 States of matter — 38
- 3.5 Freezing and melting — 40
- 3.6 Chemical reactions — 42
- 3.7 Making new products — 44
- Looking back Topic 3 — 46

Topic 4 Energy and light — 47

- 4.1 Shadow investigation — 48
- 4.2 Energy — 50
- 4.3 Energy and movement — 52
- 4.4 Energy transfer — 54
- 4.5 How light travels — 56
- 4.6 Seeing objects — 58
- 4.7 Reflecting light — 60
- **Looking back Topic 4** — 62

Topic 5 Electricity — 63

- 5.1 Wire investigation — 64
- 5.2 Why won't it work? — 66
- 5.3 Simple switches — 68
- 5.4 Series circuits — 70
- 5.5 Conductors and insulators — 72
- 5.6 Lightning conductors — 74
- **Looking back Topic 5** — 76

Topic 6 Planet Earth — 77

- 6.1 Earth's structure — 78
- 6.2 Volcanoes — 80
- 6.3 Earthquakes — 82
- 6.4 Earthquake alerts — 84
- **Looking back Topic 6** — 86

Topic 7 Earth and beyond — 87

- 7.1 Day and night — 88
- 7.2 The Earth rotates on its axis — 90
- 7.3 The Solar System — 92
- 7.4 Stars, planets, asteroids, comets — 94
- **Looking back Topic 7** — 96

Glossary — 97

Topic 1 Life processes and ecosystems

In this topic you will discover how important exercise is for the health of your body. You will also research the importance of medicine for humans, and find out that plants and animals can also get diseases.

You are going to look more closely at the habitats of animals and how they survive outside their natural environment. You will learn about food chains and how animals depend on plants and other animals for their survival. You will also learn that plants and animals need energy to survive.

1.1 Exercise for health

Key words
- exercise
- flexible
- stamina

Running, cycling and playing sports are all forms of **exercise**. Eating a balanced diet is very important if we want to be healthy, but a balanced diet without exercise will not make us strong or **flexible** or build up our **stamina**. When we exercise, energy that we get from food is used to build strength in our body instead of being stored as fat. Regular exercise can make us healthier, faster, stronger and more flexible.

To build your stamina, you need to exercise regularly. Running, swimming, dancing, or playing a sport like football or hockey are all good forms of exercise for building stamina. ▶

▲ To become more flexible, you need to stretch. Can you touch your toes like this girl can?

▲ To increase your strength, you need to do exercises like sit-ups and push-ups. These build up strength in our legs, arms and body.

1 Match each word to its meaning.

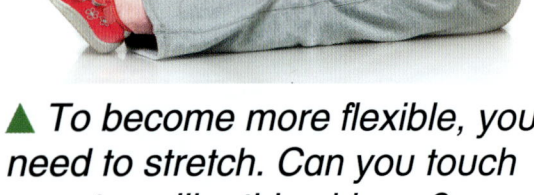

strength	being bendy or supple
flexibility	being able to carry on doing something for a long time
stamina	being powerful, able to do many things

2 Give one example of a way you can improve your strength, flexibility and stamina.

Topic **1** Life processes and ecosystems

3 Explain why exercise is important.
4 Look at the photos showing people doing a sit-up and a push-up. Explain the different steps to your partner. ▼

Doing a sit-up

1

2

3

Doing a push-up

1

2

3

Activities

1. Run on the spot for 60 seconds. How do you feel? How does your skin feel? How does your heart feel? What do you feel like in half an hour?

2. Do you know some good exercises? Demonstrate them to the class. Say if they build stamina, strength or flexibility.

3. In your group design a training programme for Stage 4 students. The programme should allow for three sessions a week. Follow your programme for a month. Record your findings.

I have learned

- Regular exercise keeps us healthy.
- Regular exercise can increase strength, flexibility and stamina.

1.2 Medicines

Key words
- drugs
- medicine
- prescription
- pharmacy

A drug is any substance that affects how your body works. For example, some **drugs** kill germs, some stop you from feeling pain and some make you sleep.

Drugs that are used to treat illness are called **medicines**. Some medicines are taken by mouth (swallowed), some are injected into the body, some are inhaled and others may be absorbed through the skin.

Doctors sometimes give sick patients a **prescription** for these medicines to treat or prevent an illness. You have to take the prescription to a **pharmacy** or a clinic to get the medicines.

Some examples of prescription medicines are:
- an inhaler to treat and control asthma
- antibiotics to treat infections caused by bacteria (germs)
- anti-retrovirals to control symptoms of HIV/AIDS.

1 Have you ever taken any medicine? What was it for?

Medicines can help you to stay healthy or to recover when you are sick. But medicines are drugs and they can be dangerous if they are not used in the correct way.

2 Some pills or tablets that are given as medicine look like sweets. Explain why you should never eat things unless you know what they are.

Topic **1** Life processes and ecosystems

Activities

1 Read the information label from a box of Nopain medicine. What information does the label give you? Why is this information important?

2 Mrs Smith is a diabetic woman. She has a headache and wants to take some Nopain medicine. Read the information on the box and tell Mrs Smith what she should do. ▶

3 Prepare a list of instructions for storing medicines safely for a family with young children.

Nopain 200 mg tablets

Indications
Rapid relief from backache, muscle pain, aching bones and joints, headache and toothache.

Dosage and directions
Adults and children 12 years or older: 2 tablets up to three times per day.
Take after meals. Swallow tablets with water.
Do not take more than 6 tablets in any 24-hour period.
Do not give to children younger than 12 years of age.
Do not take for longer than 10 days.

STORE IN A COOL DRY PLACE OUT OF THE REACH AND SIGHT OF CHILDREN.

Speak to your pharmacist/doctor before taking this medicine if you:
- have asthma, diabetes, high blood pressure or kidney problems
- smoke
- are taking any other medicines.

I have learned

- Drugs are substances that can change our bodies.
- Drugs that treat illnesses are called medicines.
- You must never take medicines unless you know what they are.

1.3 Stopping disease

All plants and animals can get **diseases**. Diseases can vary from place to place due to the seasons, the environmental conditions and the type of bacteria or virus present.

Mildew, rust and blight are common diseases that plants can get. Diseases can cause crops to rot and die.

1. Why do you think it is important for plant crops to be protected from disease?
2. How would blight on a crop of wheat affect a farmer?

Farmers can spray their crops with **pesticides** to prevent disease from spreading.

Lethal yellowing disease in coconuts is a problem for farmers in the Caribbean. The disease is very infectious and is spread by insects that transfer it from tree to tree. Scientists are still working to find a way of preventing the disease. Trees that are infected have to be destroyed.

Key words
- disease
- pesticides
- vaccination
- veterinary medicine
- quarantine

Topic 1 Life processes and ecosystems

Diseases in humans and other animals can be prevented through **vaccination**. A vaccination is usually when a very small amount of the disease is injected into our body. It is just the right amount to teach our body how to fight against the disease, but not enough to make us very ill.

3 Explain how a vaccine works.

Veterinary medicine deals with the prevention and treatment of diseases in domestic and wild animals.

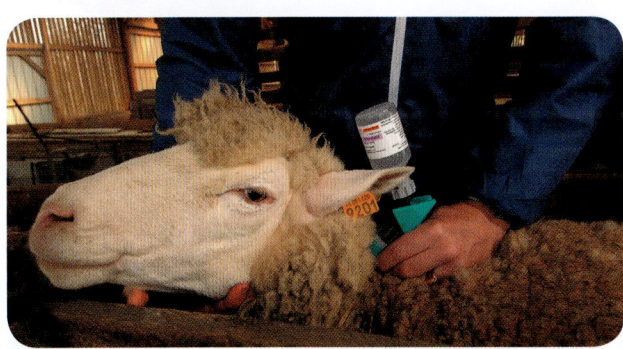

In the past, people used to kill or **quarantine** animals when they became ill with an infectious disease. Now, we try to prevent disease by vaccinating animals.

Rabies is a disease in animals that can be passed on to humans. Vets can vaccinate pets and farm animals against rabies. Some countries only allow animals in if they have been vaccinated against rabies.

4 What does 'quarantine' mean? Explain how quarantine helps to prevent disease spreading.

5 Can all diseases in animals be vaccinated against?

Activities

1 Do some research to find out how lethal yellowing disease is spread and how to prevent it. Present your research to your class. Be prepared to answer any questions they may have.

2 Do some research to find out about rabies. Answer the questions on page 6 of your Workbook.

3 Do some research to find out about anthrax in cattle. Make a fact sheet to show how the disease is spread and how to prevent it.

I have learned

- Plants and animals can contract infectious diseases.
- Vaccinations can be used to prevent some infectious diseases.

Science in context

Key word
- vaccine

1.4 Vaccines

The first **vaccine** was developed over 200 years ago. Edward Jenner used scientific enquiry to produce a vaccine that would prevent the often deadly disease called smallpox. By 1979 the smallpox vaccination had been so successful the disease was declared to be eradicated.

Jenner was a doctor in the countryside.

He observed that dairy maids who had already had cowpox did not catch smallpox.

Jenner took some cowpox virus and infected a young boy with it. Six weeks later he injected the same boy with the smallpox virus.

The boy survived and Jenner repeated his test many times on other patients to make sure it was reliable.

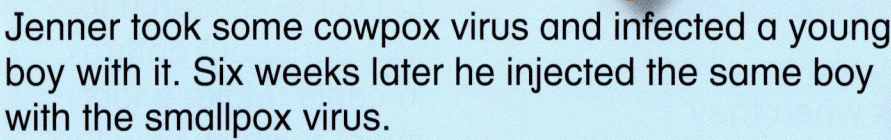

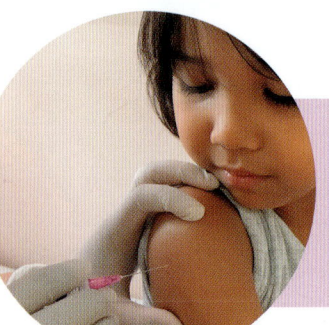

The smallpox vaccine was used across the world and now smallpox no longer exists.

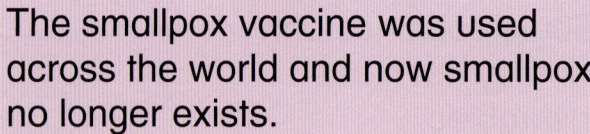

1 What pattern did Edward Jenner observe in dairy maids?
2 Describe how repeated observations give a more reliable result.

Topic 1 Life processes and ecosystems

Measles is one of the most infectious diseases in humans. It spreads very quickly and can lead to death. In 1964 a large trial was undertaken by scientists to test a measles vaccine. It worked, and four years later a vaccination programme was started. The number of measles-related deaths fell rapidly around the world.

3 Why do vaccinations need to be scientifically tested before they can be used by everyone?

The measles vaccine was later combined with a vaccine for two other diseases – mumps and rubella. It is known as the MMR vaccine.

However, in 1998 an article said the MMR vaccine was unsafe. Many people stopped having their children vaccinated and measles once again began to spread and people died. In 2004 the report was proven to be wrong. Scientists discovered that the scientific enquiry had not been done properly and that the MMR vaccine was safe.

Children in many countries are vaccinated against preventable diseases such as measles, mumps, rubella, hepatitis B, polio and diphtheria. These vaccinations have all been made possible through careful scientific enquiry.

Activities

1 Find out about childhood vaccinations in your local area. What vaccines do children have and at what age? Draw a timeline of vaccinations.

2 Do you think childhood vaccination programmes have a positive or negative effect in your local area? Research the phrase 'herd immunity' and create a dictionary definition for it.

3 Create a poster or perform a short video encouraging parents to vaccinate their children. Explain the reasons why vaccination is important.

I have learned

- Humans can be given vaccines which prevent them from catching preventable diseases.

1.5 Habitats for animals

Key words
- environment
- habitat
- adapted

The **environment** is the scientific name for our surroundings and the conditions found in them. The place in the environment where an animal (or plant) lives is called its **habitat**.

Look at these two pictures (below and opposite) carefully.

1 Describe the habitat in each picture. ▼▶
2 Make a list of all the animals that live in each habitat.

The habitat in which an animal lives must provide it with food, water and shelter. The rock-pool habitat is suitable for small sea animals such as crabs, starfish, limpets and anemones. It is also suitable for seabirds that feed on these small animals.

Many animals are **adapted** to suit their habitats. For example, the limpets are attached to the rocks by a strong muscular foot so they do not get washed off when the tide comes in or goes out. Limpets also have a strong, pyramid-shaped shell that is difficult to break, so they seldom get eaten.

Topic **1** Life processes and ecosystems

3. In what ways are the animals that you listed adapted to suit their habitats?
4. Why is the rock-pool habitat unsuitable for the rabbit and butterflies?

Scientists investigating habitats sometimes find a plant or an animal they do not recognise. They use the features of the plant or animal to find out what it is using a simple tool called a key. Keys allow scientists to group living things using their similarities and their differences.

5. Your teacher will give you a key to complete.

Activities

1. Choose one animal. Describe its habitat. Describe how it is adapted to suit the conditions found there.

2. Draw a picture of the habitat. Label it to show where the animal gets all the things it needs. Which other animals are likely to be found in the same habitat? Why?

3. Complete the activity about a pond habitat on page 9 of your Workbook. Draw a bar chart to display the information.

I have learned

- The place where an animal lives is called its habitat.
- Different animals are found in different habitats and they are adapted to suit the conditions found there.

1.6 Adapting to different habitats

Key words
- commercial
- rehabilitation

Plants and animals live and grow naturally in a specific habitat. The plants and animals thrive in their chosen area because there is enough water, food and shelter, and the climate is perfect for them.

When a plant is taken out of its natural environment, it will continue to survive as long as it has soil, water and light. However, the plant may not grow as big or be as healthy as it would if it were in its natural environment.

For example, the natural habitat of a pine tree is in the mountains where the soil is sandy and drains easily. But pine trees are grown outside of their natural habitat for **commercial** reasons. Areas where they can be easily cut down are chosen, as the wood is used for building and to make furniture.

1. Why are pine trees chosen for making furniture?
2. What makes mountainous areas unsuitable for cutting down the trees for furniture?

Topic 1   **Life processes and ecosystems**

Animals are sometimes badly injured in their natural environments. This is often due to human pollution or interference. These animals can be taken out of their natural environments and placed in an area for **rehabilitation**. Sometimes the animals are able to go back to their habitat, other times the animals do not get strong enough to return to their natural environment. The survival of these animals is dependent on the help of the trained people looking after them. The animals can be kept in safe places where people can study and learn more about them.

3 Why can a cheetah not be placed in the same natural environment as a polar bear?

4 What would a cheetah need to survive in a rehabilitation centre?

Activities

1 The leaves of a pine tree are also called pine needles. How are the leaves of pine trees adapted to cold, snowy, mountainous habitats? Design and carry out an investigation to see which shape leaves would be best for a pine tree. Record your findings on pages 11–13 of your Workbook.

2 Pine trees are adapted to getting very little water throughout the year. Why does this make them a problematic tree to have in an area that has a lot of rain throughout the year?

3 Find out about a wildlife rehabilitation or rescue centre in your local area. What types of animals do they look after? How is the environment at the centre adapted to be more like the animals' natural habitats?

I have learned

- Plants and animals can survive in environments other than their habitats.

1.7 Food chains

Living organisms in an environment interact and depend on each other in different ways. One of the ways in which organisms depend on each other is as a source of food. Both plants and animals are sources of food.

Key words
- producer
- consumer
- herbivore
- carnivore
- omnivore
- predator
- prey

1 What does a food chain show? ▼

2 In what ways do the grass, rabbit and hawk depend on each other in this environment?

When we describe what eats what in any environment, we are describing feeding relationships. The plant in this relationship is the **producer** as it makes its own food. All animals are known as **consumers** as they are dependent on plants to get their food.

We can separate the consumers into three groups: those that eat plants only are called **herbivores**, those that eat meat only are called **carnivores**, and those that eat plants and meat are called **omnivores**.

An animal that consumes another animal is known as a **predator**. The animal it eats is its **prey**.

3 Why do all food chains start with a plant?

4 What would happen to the herbivores in the grassland food chain if the carnivores were all killed? What effect would this have on the producer?

5 Name the predator and the prey in the food chain opposite. ▶

14

Topic **1** Life processes and ecosystems

bee-eater

bee

flower

Activities

1 Make a list of four wild animals. Draw food chains to show what these animals eat and what eats them (if they are eaten). Say what type of consumer each animal is.

2 Look at the word search on page 18 of your Workbook. Find the producers and consumers and draw arrows to show what is eaten by what. Answer the question.

3 Design and make a set of cards that could be used to show at least three different food chains that you would find in a tropical rainforest.

I have learned

- Food chains show the feeding relationships in an environment.
- Plants are the producers in a food chain as they make their own food.
- Animals are the consumers as they rely on the plants and/or other animals for their food.
- Consumers can be divided up into herbivores (plant eaters), carnivores (meat eaters) and omnivores (plant and meat eaters).

1.8 Energy for survival

Key words
- energy
- photosynthesis

All living things need **energy** to stay alive. Right now your body is using energy to read, to breathe and to grow. This energy comes from food. Without the energy your body gets from food, you would not be able to stay alive, to move or to grow.

The Sun is the main source of energy for all living things. Plants make their own food by using light energy from the Sun through a process called **photosynthesis**. During photosynthesis, plants trap light energy with their leaves.

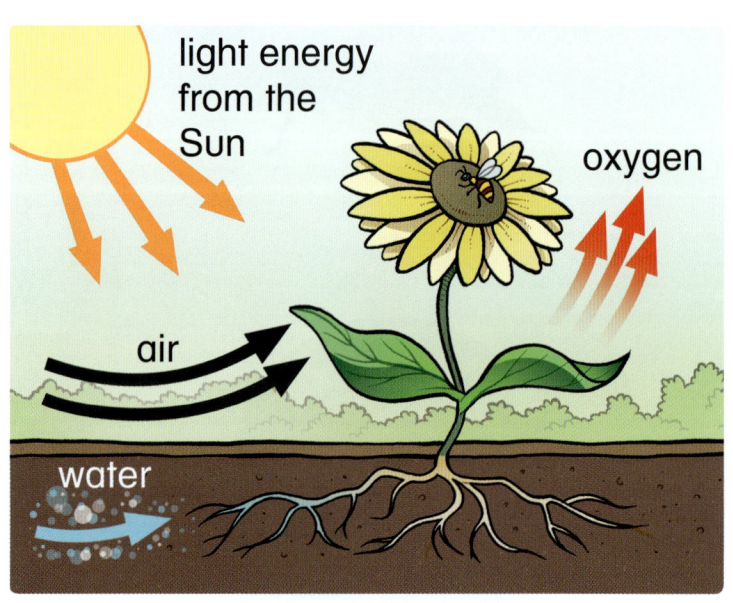

The plants use the light energy to change water and carbon dioxide into a sugar called glucose. The glucose is used by plants for energy.

1. What do you think would happen if a plant never had sunlight?

2. Look at the photographs. Why is there a difference between the leaves grown in sunlight and the leaves grown in darkness? ▼▶

A

Light keeps the leaves of a plant green and healthy.

B

This plant has not had enough light. The leaves cannot make food for the plant to grow.

Topic 1 Life processes and ecosystems

All animals get their energy indirectly from the Sun. The animals that eat plants (herbivores) get the stored energy from the leaves when they eat the plants. Animals that eat only meat (carnivores) get their energy from the herbivores they eat.

carnivore

herbivore

plant

3. Work in pairs. Tell your partner what role the Sun has as a source of energy for animals.

Humans and other animals need energy to grow, live and be healthy. All of these are important functions for survival. Your body turns the food you eat into energy.

Activities

1. Choose an animal and write a paragraph about it, explaining how it gets its energy.

2. Make a poster to show the different ways herbivores, carnivores and omnivores get their energy.

3. Investigate how humans get their energy. Research the main different food groups and fill in the food pyramid diagram your teacher will give you.

I have learned

- All living things get their energy from the Sun.
- Plants trap light energy to help them make their own food.
- Animals get their energy from the meat and/or plants they eat.

17

Looking back Topic 1

In this topic you have learned

- Some drugs can be used as medicines to treat, control or prevent illnesses.
- Plants and animals can contract infectious diseases. Animals can be vaccinated against some infectious diseases.
- The environment is the scientific name for our surroundings and the conditions found there. The area in which an animal lives is called its habitat. There are many different habitats in an environment.
- Plants and animals are adapted to their environment.
- Food chains always start with the producer, which is always a plant as plants make their own food.
- Animals are the consumers in a food chain as they eat either plants or other animals to survive.
- Animals can be classified according to what they eat: herbivore (eats plants), carnivore (eats other animals), omnivore (eats plants and animals).
- Plants and animals need energy to grow live and be healthy.

How well do you remember?

1 Why is exercise important for humans?

2 Complete the sentence. *When a plant is taken out of its natural environment, it will continue to survive as long as it has* _____, _____ *and* _____.

3 Why should you never take medicines that were prescribed for someone else?

4 Name an animal that lives in a desert habitat.

5 Classify the animals in this food chain according to what they eat.

algae → shrimp → turtle → shark

Topic 2 — Humans and other animals

In this topic you are going to learn more about how models are used in science to help understand science better. You will focus on the human body, looking at the human skeleton, how bones move and the functions of the skeleton. You will also look at animal skeletons and how they can differ from human skeletons.

Thinking and working scientifically

2.1 Modelling the human body

Key words
- model
- simplified

Models are used by scientists to help explain something more easily. They are often **simplified** diagrams of a real thing.

Models are not always completely accurate representations of the real world, but as understanding and knowledge develops around what scientists are researching, the representation will become more accurate.

1. How does the diagram of the human skeleton we use today compare with the one drawn by the Greek doctor Galen 1900 years ago? ▼ ▶

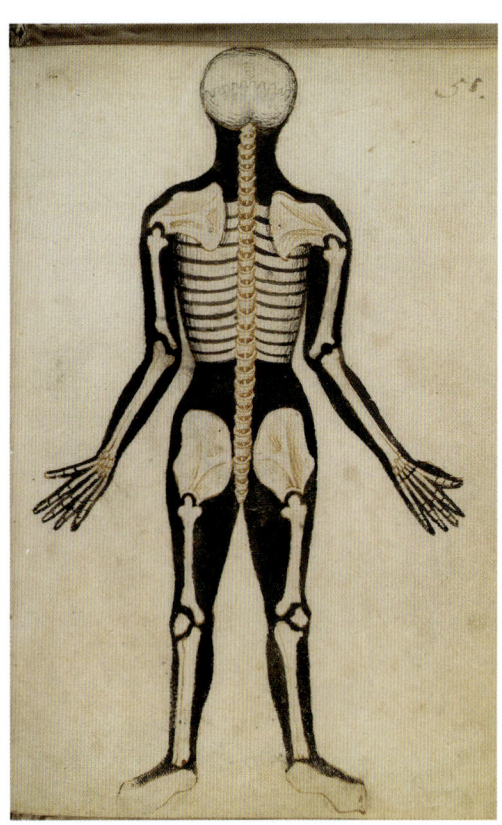

Galen's diagram

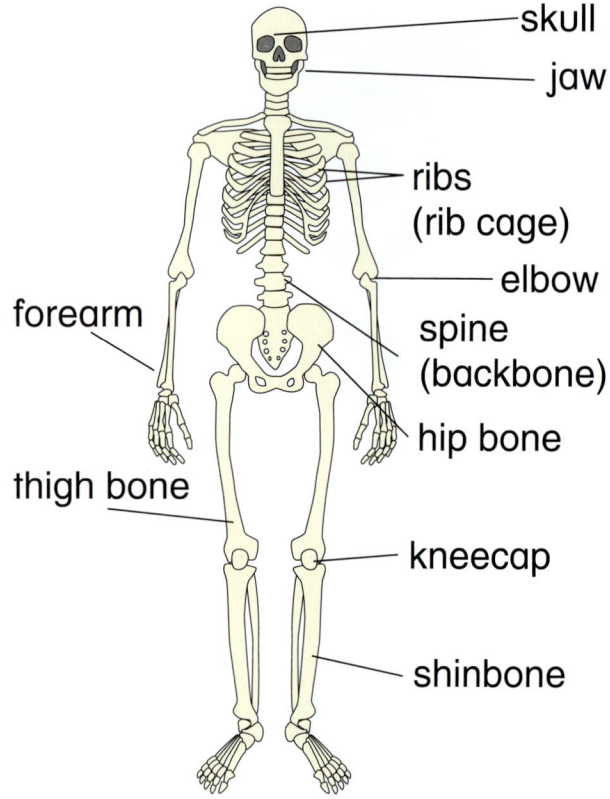

A modern diagram

Models can be used to help us understand the relationship between different things.

Topic **2** Humans and other animals

2 How does this diagram of arm muscles contracting and relaxing help us to understand the relationship between the muscles and bones in our body? ▼

Biceps contracted, triceps relaxed (extended)

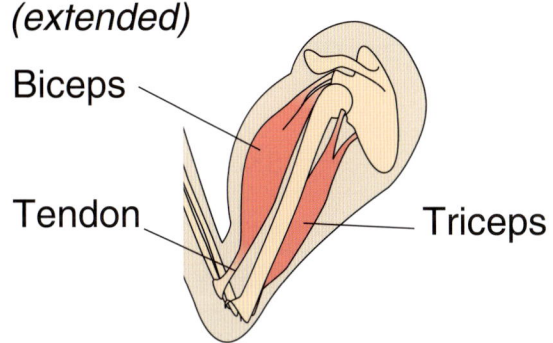

Biceps
Tendon
Triceps

Triceps contracted, biceps relaxed

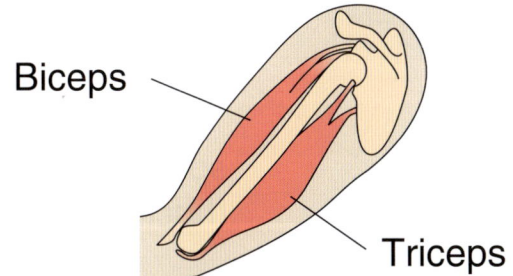

Biceps
Triceps

Models can also show quantities by showing the number of something accurately.

Models can be smaller or larger representations of a real thing, but they are made or drawn to scale to make the model more realistic.

3 Look at the three-dimensional human skeleton model your teacher has. Explain how it shows quantity.

4 Why is it important for a model to be made or drawn to scale?

Activities

1 Draw a diagram to show what happens to a muscle when it contracts.

2 Investigate whether you could use the model of the arm to show how your leg bends at the knee. Explain why or why not.

3 The muscles attached to your skeleton are voluntary muscles. This means that you decide whether to move the muscles or not. Some of the muscles in your body are involuntary muscles. Find out what this means.

I have learned

- Models are not fully representative of a real-world situation or scientific idea.
- Models are used to show relationships, quantities or scale.

2.2 The human skeleton

Key words
- skeleton
- joints

Like some animals, human beings also have a bony skeleton inside their bodies.

When you squeeze your arm you can feel the bones inside it. Bones are hard and strong and they form a rigid frame inside your body. This frame of bones is called a **skeleton**. Your skeleton helps you to move and supports your body, as well as protecting the soft parts inside your body.

1. Do you know the names of any of the bones in your body? Show them to your partner and say their names.

There are 206 bones in a human skeleton. On their own, bones are hard and stiff and they cannot bend. However, the bones in our skeletons are connected to each other at **joints**. Because joints are flexible, we are able to bend and move our bodies.

2. Find three joints on the skeleton. In what way does each joint allow you to move your body? ▶

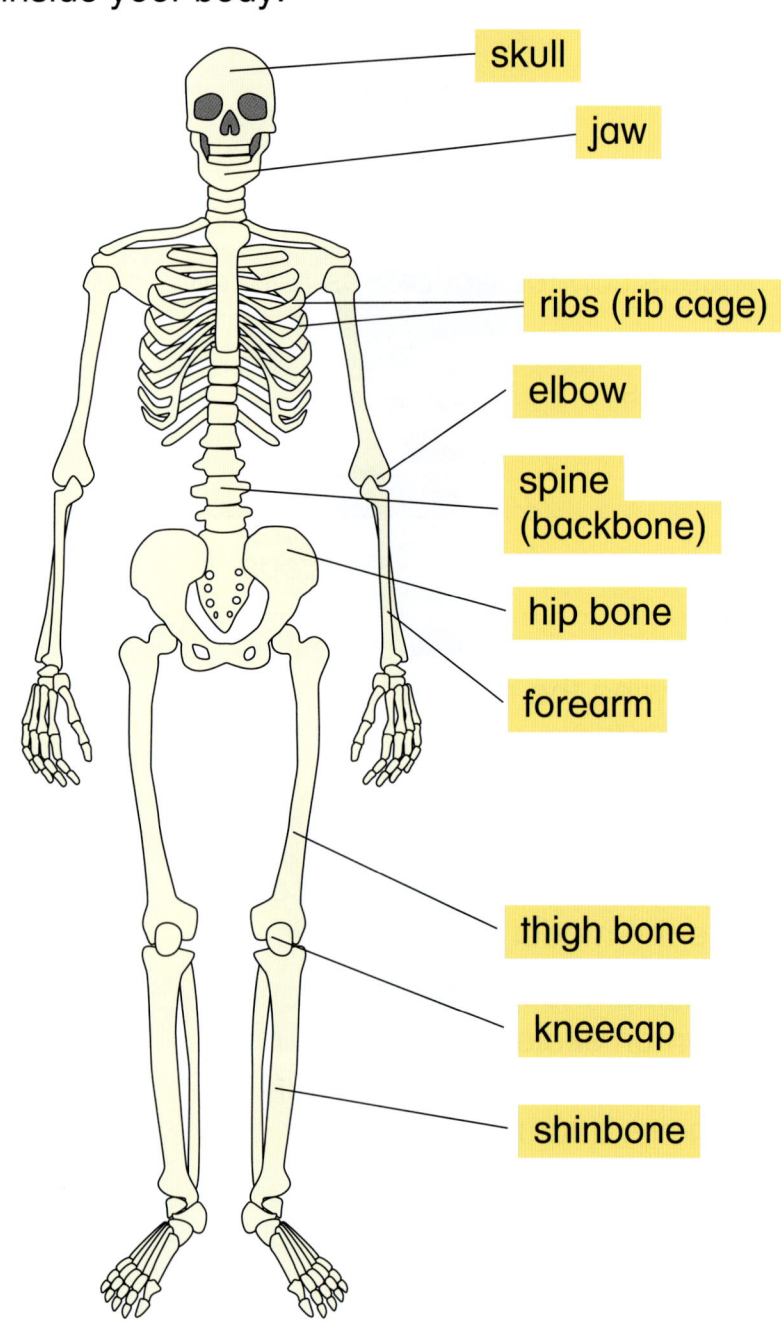

- skull
- jaw
- ribs (rib cage)
- elbow
- spine (backbone)
- hip bone
- forearm
- thigh bone
- kneecap
- shinbone

22

Topic 2 Humans and other animals

Bones in museums and the bones of dead animals in the environment are dry and they crumble easily. Living bones are not dry and are very strong.

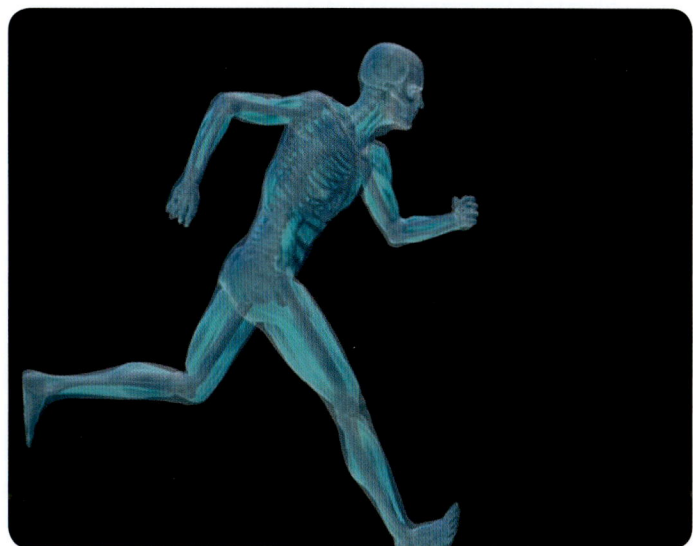

Activities

1. Describe some differences between the bones that make up the human skeleton.

2. Find out what happens when you break (fracture) a bone. What is the difference between a complete fracture, a greenstick fracture and an open fracture?

3. When you break a bone you may have to wear a splint or a cast. Make a poster to teach young children what they can expect if they have to wear a splint or a cast.

I have learned

- Humans have bony skeletons inside their bodies.
- Skeletons are made of different kinds of bones connected to each other at joints.

2.3 Moving your bones

Key words
- muscles
- contract
- relax

The skeleton and **muscles** work together so that you can move.

1 What are the muscles around our skeleton attached to?

Muscles move the bones they are attached to by **contracting**, or shortening. When a muscle contracts, it pulls against the bone and the bone moves. Muscles cannot push – they can only pull – so they need to work in pairs to allow each joint to move.

The elbow is an example of a joint. It is controlled by one set of muscles at the front of the upper arm and another set of muscles at the back of the arm.

Look at the pictures and read the information to see how one pair of muscles works together to allow you to bend and straighten your arm.

When you bend your arm:
- The muscles at the front of the upper arm get fatter, harder and shorter.
- The muscles at the back of the upper arm get thinner, softer and longer (they **relax**).
- The lower arm moves up.

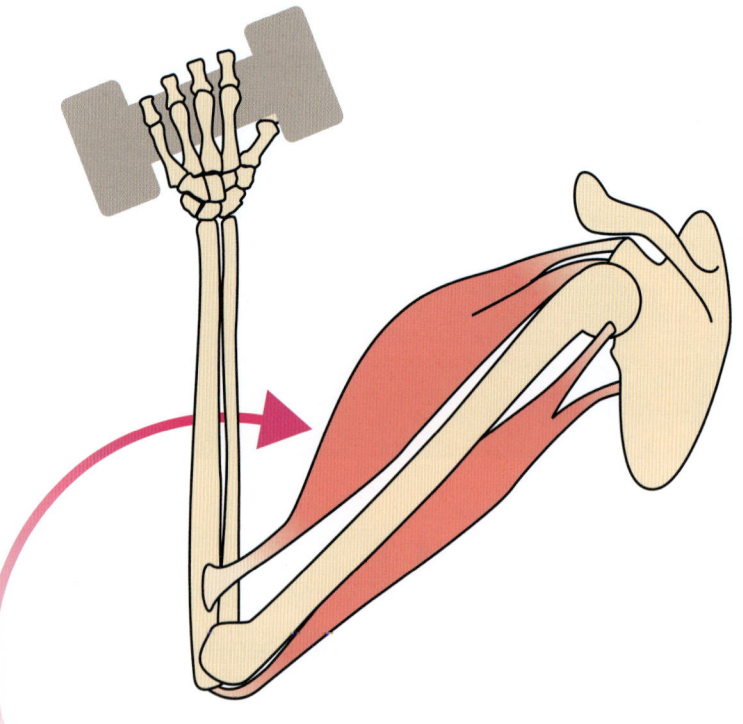

Topic 2 — Humans and other animals

When you straighten your arm:

- The muscles at the front of the upper arm get thinner, softer and longer (they relax).
- The muscles at the back of the upper arm get fatter, harder and shorter.
- The lower arm moves down.

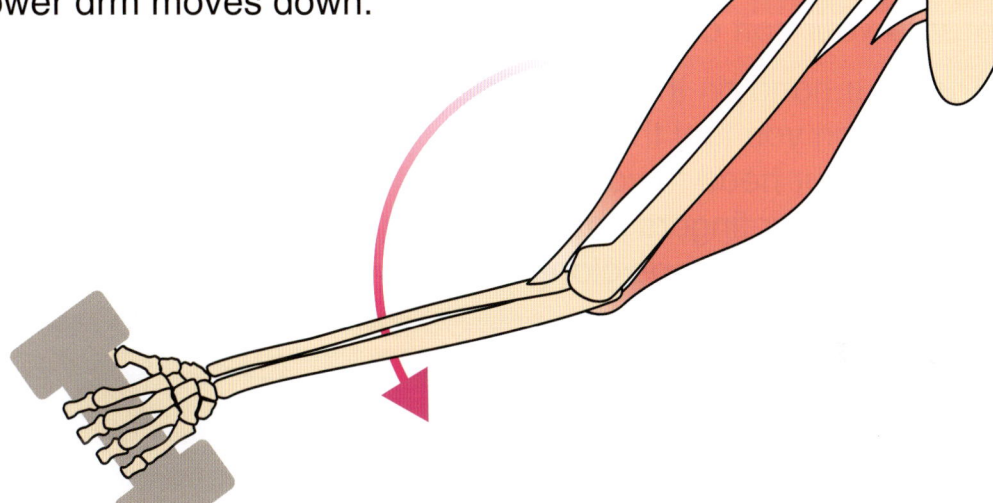

Activities

1. Draw a diagram to show what happens to a muscle when it contracts.

2. Test out the muscles in your legs. Explain to a partner which muscle is contracting and which one is relaxing when you:
   - lift and lower your leg
   - bend your leg at the knee.

3. Write a short paragraph in your notebook about why the skeleton and muscles need to work together as a system.

I have learned

- Muscles have to shorten to make a bone move.
- Muscles act in pairs around joints to allow us to move.

2.4 Functions of the skeleton

Key words
- support
- protect
- organs
- backbone

Without your skeleton you would not be able to stand, walk or run. Your body would be soft and shapeless. Your skeleton allows you to stand upright and to move. It also **supports** your body and **protects** your soft **organs**.

1. Put your hand on the back of your neck. Can you feel the bones joining your head to your body? Now run your fingers downwards. Try to feel the bumps along your **backbone**.

Your backbone, or spine, is important for supporting your body. It also allows you to twist and bend.

2. What do you think the discs of cartilage do?

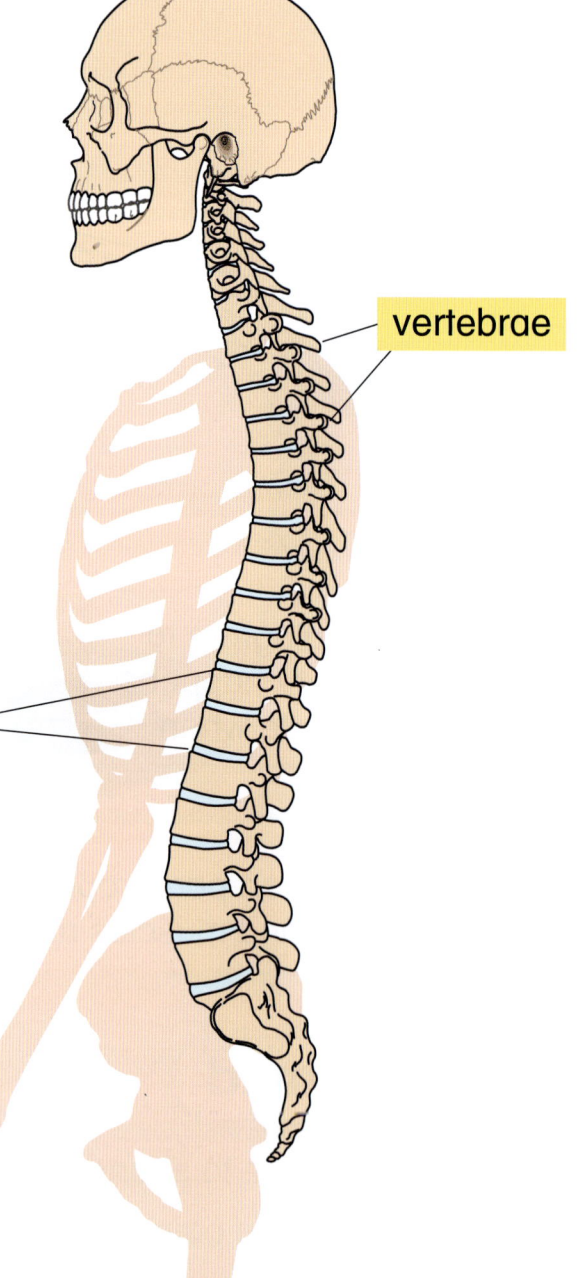

vertebrae

discs of rubbery cartilage

Your skeleton protects the soft organs inside your body.

Your skull protects your brain.

Your heart, lungs and liver are all inside your chest. These important organs are protected by your ribs. The ribs form a bony cage around your organs. This is called your ribcage.

Topic 2 Humans and other animals

Your backbone is a column of small, ring-shaped bones called vertebrae (vert-eh-bray). A single one of these bones is called a vertebra. The vertebrae are separated by small cushions of rubbery cartilage. The cartilage stops the vertebrae from rubbing together and acts like a shock absorber to cushion the vertebrae from jolts when you walk and run.

Your backbone also protects the nerves that send messages from your brain to the rest of your body.

Your legs are important for support too. The leg bones are large and strong. The bone in your thigh is the largest bone in your body.

Activities

1. Make a model of a thigh bone. Decide how you will make it light but strong, like a real bone. Use pages 25 and 26 in your Workbook to record your work.

2. What can you do to test the strength of the model bones? Work with another group to test the model bones. Whose bone was stronger?

3. The backbone is different from the thigh bone. Think about how you could build a model to show the backbone and the way it allows the body to move. Draw a labelled diagram of your proposed model.

I have learned

- The skeleton allows us to move.
- It also supports our body and protects our organs.

2.5 Animal skeletons

Key words
- vertebrates
- invertebrates
- exoskeleton

Humans and other animals that have a backbone (spine) are called **vertebrates**. Animals that do not have a backbone are called **invertebrates**.

Mammals, fish, reptiles, amphibians and birds all have a backbone so they are classified as vertebrates.

frog

bird

fish

snake

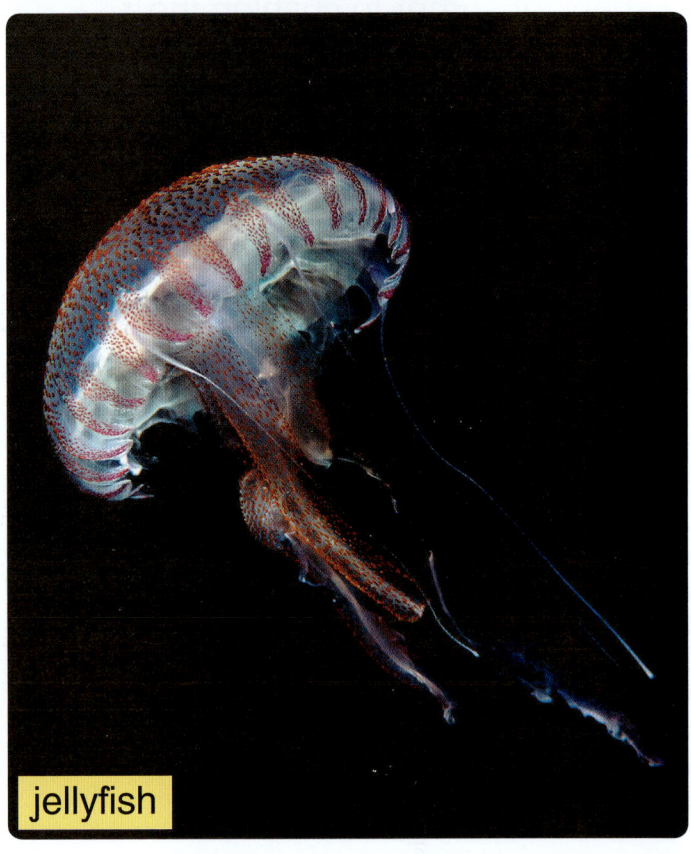
jellyfish

Look at this photograph of a jellyfish. Jellyfish have soft floppy bodies with no bones, so they are invertebrates. Other invertebrates include squid, octopuses, slugs, spiders and earthworms.

1. Can you think of any other animals with soft bodies and no bones?

Some animals have a hard covering on the outside of their body. This is called an **exoskeleton**.

Topic 2   Humans and other animals

2. Some insects, such as locusts and beetles, have a hard exoskeleton. Do you think they have bones?

Activities

1. Draw a table of vertebrates and invertebrates. Try to find at least five examples of animals in each group.

2. Make a list of at least five animals with an exoskeleton. Try to find examples that are different to those given here and in the Workbook.

3. What is an x-ray image? What are these used for in modern medicine? Find an example where they are used in veterinary medicine as well.

I have learned

- Animals with a backbone are called vertebrates. Animals without a backbone are called invertebrates.
- Some animals have an exoskeleton.

Looking back Topic 2

In this topic you have learned

- Some animals including humans have bony skeletons inside their bodies.
- Skeletons are made of different kinds of bones connected to each other at joints.
- Our skeleton allows us to move, supports our body and protects our organs.
- The bones of our skeleton have muscles attached to them. Muscles allow us to move our bodies. Muscles have to shorten to make a bone move.
- Muscles work in pairs around joints to allow movement. One muscle contracts while the other muscle relaxes.
- Animals with a backbone are called vertebrates. Animals without a backbone are called invertebrates. Some animals have an exoskeleton.

How well do you remember?

1 Look at the diagram of an elephant.
   a What is the animal's skeleton made of?
   b Name one animal that does not have a skeleton.
   c Write down two similarities and two differences between an elephant's skeleton and a human skeleton.

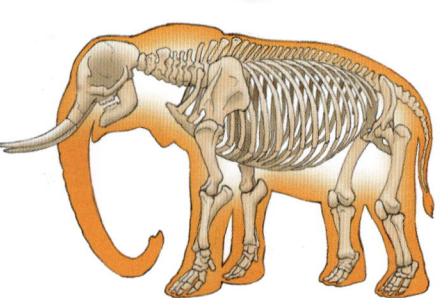

2 What are two advantages of having a bony skeleton inside your body?
3 Name three animals that have an exoskeleton.
4 True or false? Muscles working in pairs always pull and never push.

Topic 3 States of matter

In this topic you are going to learn more about matter and materials. You will examine solids, liquids and gases. You will work with materials to investigate heating and cooling further. You will focus on water to learn more about melting, solidifying, freezing and what happens to steam when you cool it.

Thinking and working scientifically

3.1 Investigating play slime

> **Key words**
> - scientific enquiry
> - investigate
> - prediction
> - compress

In Stage 3, you learned that **scientific enquiry** is about investigating a question and that there are five different types of scientific enquiry: research, fair testing, observing over time, identifying and classifying, and pattern seeking.

To carry out a scientific enquiry, you need to ask scientific questions that can be **investigated**.

Consider the following enquiry.

You are going to investigate some 'play slime' to identify its state of matter.

1 Make a **prediction** about what you think play slime is. Is it a solid, a liquid or a gas?

2 Do some research to find out how you could answer the following questions about play slime.

  a Can it be **compressed**?

  b Does its shape change according to the container it is in?

  c Does it always take up the same amount of space?

Topic **3** States of matter

3 How will you make this a fair test? What will stay the same and what will change?

4 What will you do to find the answers to the questions you asked?

5 How will you observe what happens?

6 Is there anything you need to wear or do to make sure you cannot get hurt during the investigation?

Now do your investigation.

7 Write down what happened during your investigation.

8 Can you classify 'play slime'? Use the table below to help you.

	Solid	Liquid	Gas
Can it be compressed?	No	Yes	Yes
Does its shape change according to the container it is in?	No	Yes	Yes
Does it always take up the same amount of space	No	Yes	No

Activities

1 Which types of scientific enquiry did you use in the investigation?

2 Why is it important to make sure the investigation is a fair test?

3 What safety measures need to be thought about during various investigations? Make a poster about safety in the science lab.

I have learned

- The five main types of scientific enquiry are research, fair testing, observing over time, identifying and classifying, and pattern seeking.

3.2 Materials, substances and particles

Matter is the scientific word for what everything is made of. Matter is found in three different states: solid, liquid and gas.

Materials are made from matter. Metals, plastics, ceramics, glass and fabrics are examples of materials.

Key words
- material
- substance
- composition
- element
- compound
- particle
- particle model

1. Look at these items made from different materials. Identify the materials used in each. ▼▶

Materials have different properties. For example, some are hard, others are soft; some are strong while others are weak. Materials can also be shiny or dull.

2. List the properties of the materials in the pictures.

A **substance** is matter which has specific properties and **composition**. There are two types of substances: **elements** and **compounds**. An element is a substance in its simplest form, for example gold. Compounds are made from two or more elements, for example water is a combination of hydrogen and oxygen.

Topic 3 States of matter

3 Can you name five substances around you?

All substances are made of **particles**, which are the smallest parts of substances. Each substance is made up of unique particles. A **particle model** helps scientists to explain the differences between solids, liquids and gases.

4 Look at the particle models of a solid, liquid and gas. How are they different? ▼

Solids	
The particles in solids are very close together and arranged in order.	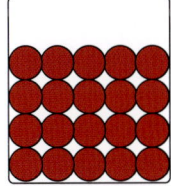
Liquids	
The particles in liquids are not so close together. What do you notice about the order?	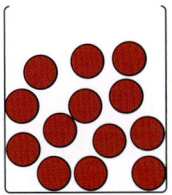
Gases	
The particles in gases have large spaces between them. What do you notice about the order?	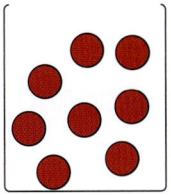

Activities

1 Think about how you can tell the difference between a material, a substance and a particle. Write down your ideas.

2 Test some materials to find out about their properties. Use pages 30–32 of your Workbook to record your findings.

3 Draw up a table to compare materials and substances. Include drawings of each one.

I have learned

- Materials are made of one or more substances.
- Substances are either elements or compounds.
- Particles are the smallest parts of a substance.

3.3 Solids and liquids

Key words
- particle model
- vibrating
- regular
- random

You already know that particles are the smallest parts of a substance and are too small for us to see. Scientists use the **particle model** to help explain the differences between solids, liquids and gases.

Scientists have identified that particles are always moving and **vibrating**, even in a solid. The particles are also attracted to each other, although there are spaces between them.

The arrangement of the particles in a particle model will depend on whether the substance is a solid, a liquid or a gas.

1. Look at the particle model of a solid. What can you tell about a solid by looking at the particle model? ▶

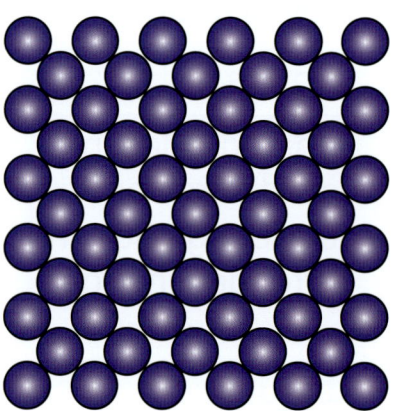

You can see that the particles of a solid are packed closely together and are arranged in a **regular** way. Although the particles are vibrating, they cannot move around from place to place.

2. Look at the particle model of a liquid. What can you tell about a liquid by looking at the particle model? ▶

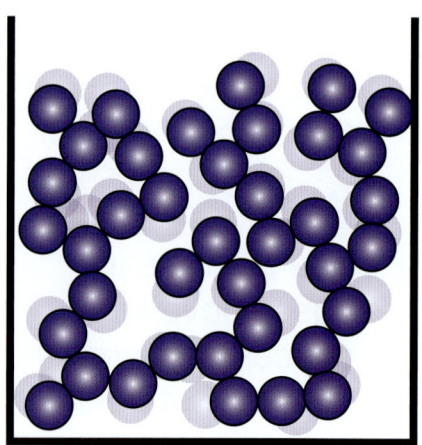

You can see that the particles of a liquid fill the shape of the container they are in and that each particle touches at least one other particle. They are arranged in a **random** way and can move around each other. The particles cannot be squashed.

3. Look at the picture of sugar. Is sugar a solid or a liquid? Why? ▼

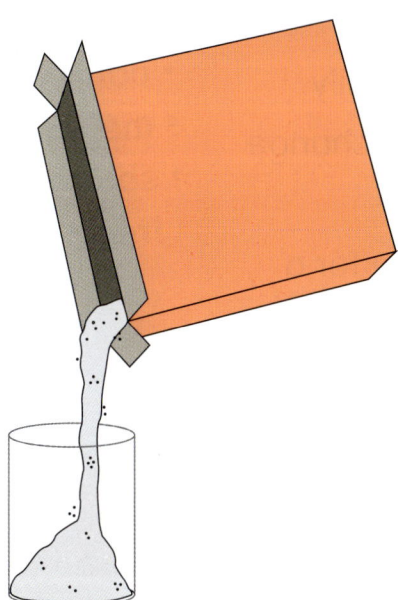

Some solids can be confused for liquids as they can be poured into a container. Powders are crushed solids and have not changed their state. One grain of sugar will not spread out on the bottom of a container, but will keep its shape.

4. Give three more examples of powder solids which could be confused for liquids.

Activities

1. Think about how you can tell whether matter is in a solid or liquid state. Write down your ideas.

2. Write a list of the solid materials in your classroom. How do you know these are all solids?

3. Draw up a table to compare the properties of a powder solid with a liquid. Include drawings of the particle models to help with your explanation.

I have learned

- The particles of a solid are packed closely together and are arranged in a regular way. They cannot move from place to place.
- The particles of a liquid fill the shape of the container they are in and each particle touches at least one other particle. They are arranged in a random way and can move around each other.

3.4 States of matter

Key words
- physical changes
- reversible
- evaporation
- condensation
- melting
- solidifying
- freezing

Changes of state are **physical changes** in matter. These changes are **reversible** changes. This means that the matter is not changed chemically.

Changes in temperature will cause matter to change between the different states.

1. What needs to happen to the temperature for a liquid to turn to a gas?

When temperature is increased to matter in a liquid state, it will change from a liquid to a gas. This process is called **evaporation**. To reverse this process, the temperature needs to be lowered, and the matter will return to its liquid state. This process is called **condensation**.

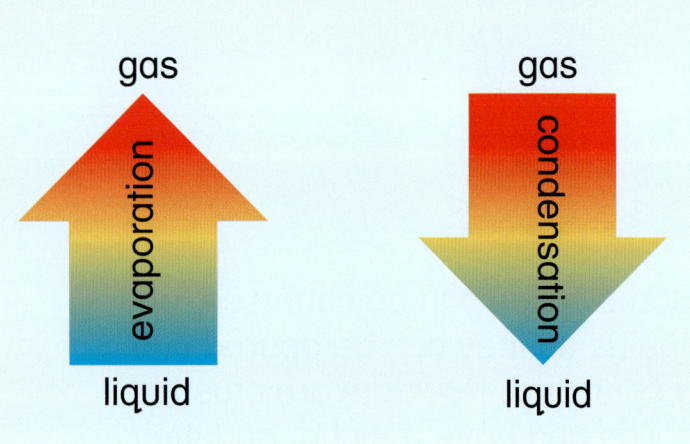

2. What needs to happen to the temperature for a liquid to turn to a solid?

When temperature is increased to matter in a solid state, it will change from a solid to a liquid. This process is called **melting**. To reverse this process, the temperature needs to be lowered, and the matter will return to its solid state. This process is called **solidifying** or **freezing**.

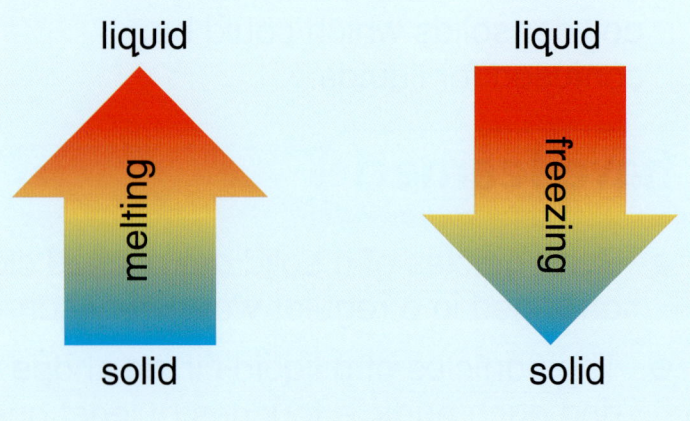

Topic **3** States of matter

3 Draw particle models to show the changes in the states of matter.

The pictures below show some different changes of state.

4 Identify the changes in the states of matter.

Activities

1 Imagine you are a water droplet in an ice block. Draw a cartoon story to show what happens to you as you move from a solid to a liquid to a gas.

2 Draw diagrams to show the physical changes of the state of water. Follow the instructions on pages 36–37 of your Workbook.

3 Find out how heating is used to extract precious metals, such as gold and silver, out of the rocks in which it is found. Draw a labelled diagram to show how this is done.

I have learned

- Matter can change between the different states. These changes are physical changes which are reversible.

3.5 Freezing and melting

When you add **heat** to solids, such as ice or butter, they get soft and turn to liquid. This process is called **melting**.

Some solids need more heat than others to make them melt. For example, if you melt butter in a metal pan, the butter will melt but the metal pan will not melt.

Key words
- heat
- melting
- freezing
- solidify
- cooling

1 Do you think all ice melts at the same rate?

2 If you heated ice, butter and candle wax in three identical pans, over the same amount of heat, which one do you think would melt first? Why?

When liquid is cooled enough, it turns to a solid. It solidifies or freezes. To **solidify** or to freeze a liquid is the reverse of melting it. In science, **freezing** means the same as to solidify, so freezing is not just a term used for water turning to ice.

3 Do you think all liquids solidify at the same temperature?

Solidifying (or freezing) and melting are the reverse changes to the states of liquid and solid matter. This means that if you melt a solid by heating it, you can reverse the change of state by **cooling** the liquid and causing it to solidify or freeze.

Topic **3** States of matter

Look at the pictures.

4 Write about what is happening in the pictures. ▼

5 Can you think of some other examples when the processes of melting and solidifying or freezing are useful to us?

gold ore (solid)

melted gold (liquid)

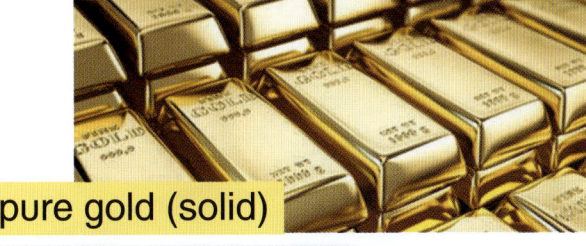
pure gold (solid)

Activities

1 Have a class competition to see which group can make an ice cube melt the fastest. Record your results on page 39 of your Workbook.

2 Investigate which solid melts the fastest by putting samples in foil dishes and floating them in hot water. Use page 40 of your Workbook to record your findings.

3 Find out what the terms 'freezing point' and 'melting point' mean. Compare the freezing point of water with that of candle wax. Explain why the melting point is actually the same as the freezing point in most matter.

I have learned

- When a solid is heated it becomes a liquid. We say it has melted.
- Some solids need more heat than others to melt.
- Cooling liquids causes them to turn to solids. This is called solidifying or freezing.
- Solidifying or freezing is the reverse of melting.

3.6 Chemical reactions

A **chemical reaction** is also called a chemical change. When substances are joined together they change to make something new. This happens when the particles are rearranged and so a new **product** is made.

Chemical changes are different to physical changes as they cannot be reversed, we say the changes are **irreversible**.

Key words
- chemical reaction
- product
- irreversible

1. Use the pictures to explain why we say chemical reactions are irreversible. ▲
2. Give three more examples of chemical reactions.

There is a chemical reaction which takes place in our bodies. As we eat food, the particles in the food are rearranged, changing the food into energy which our bodies can use. The food mixes with chemicals in our bodies and changes.

Topic **3** States of matter

Look at these pictures. ▼

3 Describe what has happened to the substances to make a new substance.

4 Do you think the processes can be reversed? Why?

Activities

1 Make a baking soda volcano. Record your observations on page 42 of your Workbook.

2 Make a bath fizzer. Record your observations on page 43 of your Workbook.

3 Make plastic from milk. Record your observations on page 44 of your Workbook.

I have learned

- Chemical reactions take place when substances are joined together to make a new product.
- Chemical reactions are irreversible.

Science in context

3.7 Making new products

Key words
- plaster of Paris
- plastic

When substances are combined in a chemical reaction, new products with different properties are produced. **Plaster of Paris** is an example of a new product which is made from a chemical reaction.

Plaster of Paris (also known as gypsum) is a very useful product as it is used when someone breaks a bone. A doctor will put the broken bones in a cast of plaster of Paris.

To make a cast, plaster of Paris powder is mixed with water and a chemical reaction takes place. Bandages are then dipped into the paste and wrapped around the bones which are broken. The paste sets hard.

1 Is it possible to reuse plaster of Paris after it has set? Explain your answer.

2 Look at the picture. Why is plaster of Paris so useful for making casts for broken bones? ▶

44

Topic **3**   States of matter

Plaster of Paris is also used in building construction on wall surfaces and by artists for sculptures.

3 What other uses could plaster of Paris have?

Plastic is another useful product made from a chemical reaction. Oil is changed during the process and cannot be turned back to oil once the new product, plastic, is made. Plastic is a useful material for creating many items.

4 Look at the pictures and identify the uses of plastic. ▼

Activities

1 Talk about the items in the pictures. Can you identify the product they are made from?

2 Why are plastic bottles often used instead of glass bottles? Which material is more useful for making bottles – plastic or glass? Explain your answer.

3 Research an interesting product created by a chemical reaction and make notes. Be prepared to tell the class about your interesting product.

I have learned

- Many useful products are created from a chemical reaction.
- Two examples of products created from a chemical reaction are plaster of Paris and plastic.

Looking back Topic 3

In this topic you have learned

- Materials are made of one or more substances. Substances are either elements or compounds. Particles are the smallest parts of a substance.
- The particles of a solid are packed closely together and are arranged in regular way. They cannot move from place to place.
- The particles of a liquid fill the shape of the container they are in, and each particle touches at least one other particle. They are arranged in a random way and can move around each other.
- Substances can change from a liquid to a solid, or from a solid to a liquid. These are physical, reversible changes. Examples are melting and solidifying or freezing.
- Substances which react with each other to produce a new product undergo an irreversible change. This is known as a chemical reaction.

How well do you remember?

1 Choose the correct scientific word for each statement.

   freeze   motion   reverse   solid   melt   particle

   a This is a material with a fixed shape. It does not flow. _____
   b Melting is the _____ of freezing.
   c The smallest part of a substance is known as a _____.
   d When a solid such as chocolate is heated, it will _____.
   e To change a liquid into a solid, you need to _____ it.
   f Particles are always in constant _____.

2 Explain how some solids behave like liquids.

3 Which of these is an example of a physical reaction?

A
B

Topic 4 Energy and light

In this topic you will learn more about energy and light. You will investigate energy and how it is transferred. You will identify that in the process of energy transfer, some of the energy becomes wasted energy, which can be in the form of light, heat and sound. You will work with light to understand that it only travels in straight lines. You will look at how light can reflect off surfaces and how you can see objects which are not light sources.

Thinking and working scientifically

4.1 Shadow investigation

Key words
- fair
- safe

When you carry out a scientific enquiry it is important to work slowly and carefully to make sure nothing is left out and the investigation is **fair** and **safe**.

Investigate the way shadows change during the day. Your teacher will give you some instructions.

1. What do you predict will happen in this investigation?
2. What equipment do you need to conduct the investigation? What will you use to make a shadow? How will you record the shadows?
3. How often will you observe the shadows?

4. How will observing the shadow often help you with the investigation?

Topic 4 Energy and light

5. What measurements will you need to take during the investigation? How will you take these measurements?

6. What safety precautions do you need to follow when conducting the investigation?

7. Create a table like the one below to record your measurements.

Time of day	Length of shadow	Direction of shadow (draw an arrow)

8. Use the results from the table to draw a dot plot of time of day and length of shadow.

9. Describe any patterns you can see in your results.

Activities

1. Draw a diagram to show the equipment used in the investigation. Explain your choice of equipment.

2. Draw a diagram to show what you observed over time. Write a brief explanation of your observations.

3. Using a torch and an object in class to create a shadow, recreate the investigation inside. Explain what happened in the investigation.

I have learned

- Making investigations fair and safe is very important.
- We use standard measurements to make investigations meaningful and fair.

4.2 Energy

Key words
- energy
- work
- movement energy
- stored energy
- potential energy
- transferred
- wasted energy

Everything we do uses **energy** and all **work** requires energy. Energy is present in all matter. Think back to the particle models you looked at in Topic 3. In all three states of matter the particles are moving. This is called internal energy and is a form of **movement energy**.

Other forms of energy are heat, light, sound and electric energy.

Some objects have **stored energy**, also known as **potential energy**.

Energy cannot be used up, lost or destroyed, but it can be changed from one form to another. We say the energy has been **transferred**.

Look at the flashlight. ▶ The stored energy in the cells can be transferred into electrical energy when the flashlight is switched on. The electrical energy is then transferred into light energy.

1 Describe the transformation of energy in each of these examples. ▼

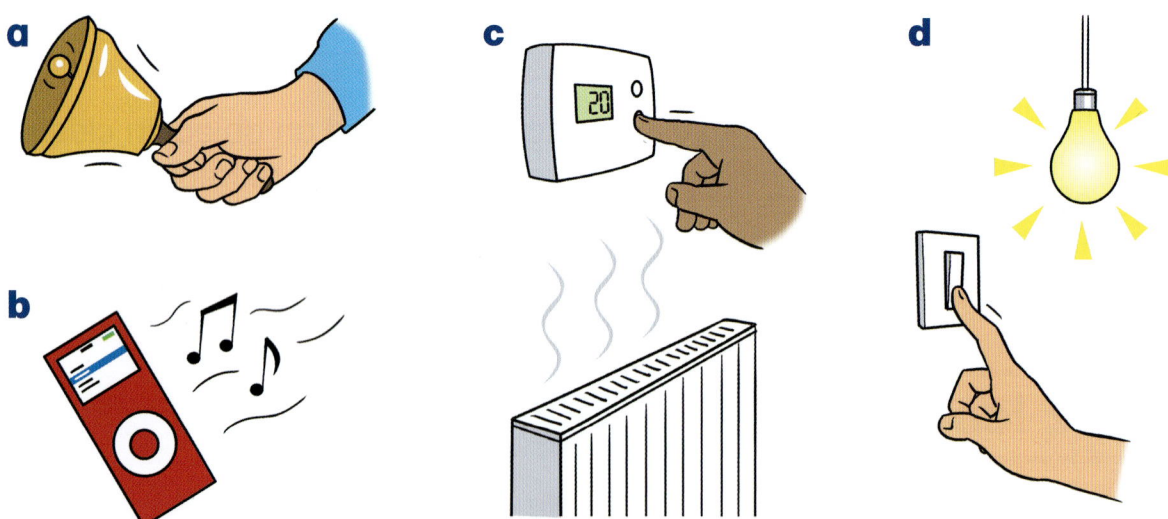

Heat, sound and light energy are examples of energy transfers.

Energy transfers can be shown on a transfer diagram.

Topic **4** Energy and light

Not all energy transferred from one object to another comes out as useful energy. For example, when we turn on a lamp we only want light energy, but some energy is transferred into heat energy. We say this is **wasted energy**.

Electrical energy goes in → Light energy comes out

Electrical energy goes in → Sound energy comes out

Electrical energy → Light energy / Heat energy

2 What form of energy goes into a television?
What forms of useful energy come out?
What form of wasted energy comes out?

Activities

1 Look at the pictures on page 47 of your Workbook. Complete the table to identify the energy in each picture.

2 Use page 48 of your Workbook to create an energy transformation diagram for an electric kettle.

3 Children running and jumping are changing stored energy from the food they have eaten into movement energy. What other things do you use the energy you store in your body for? Write a paragraph.

I have learned

- Energy is present in all matter.
- Energy cannot be used up, lost or destroyed, but it can be transferred.

51

4.3 Energy and movement

Key words
- energy
- movement

Energy allows objects to do work. In scientific language, work means more than just physical labour – it includes things such as breathing, moving, cooking food and objects falling to the ground as a result of gravity.

Look at these pictures. They all show energy making things work.

1. What work is happening in each picture?
2. Where is the energy coming from to do the work?

Topic **4** Energy and light

We need energy to move our bodies just as other objects need energy (often in the form of a force) to start them moving. In addition, all moving objects have movement energy, which means they can do work as a result of the movement.

Think about kicking a ball. Your foot moves to make the kick. The energy from your moving foot provides the force needed to start the ball rolling. The ball then has its own movement energy that allows it to do the work of rolling.

3 Make a list of six things (other than your body) that have movement energy.

Activities

1 Your teacher will give you some pictures of some athletes. Complete the table to identify the movement energy in each sport.

2 Make a poster to show that movement is a form of energy.

3 Make a model bungee jump and use it to investigate the energy in movement. Use Workbook pages 49 and 50 to record your work.

I have learned

- Movement requires energy.
- All moving objects have movement energy.

53

4.4 Energy transfer

Key words
- wasted energy
- energy saver

Energy does not disappear, but some energy can be transferred to the surrounding environment. This is **wasted energy**. When something happens, all the energy at the start is still there at the end. The output energy is equal to the input energy. This can be shown in an energy transfer diagram.

1 How do these diagrams compare amounts of energy? ▼

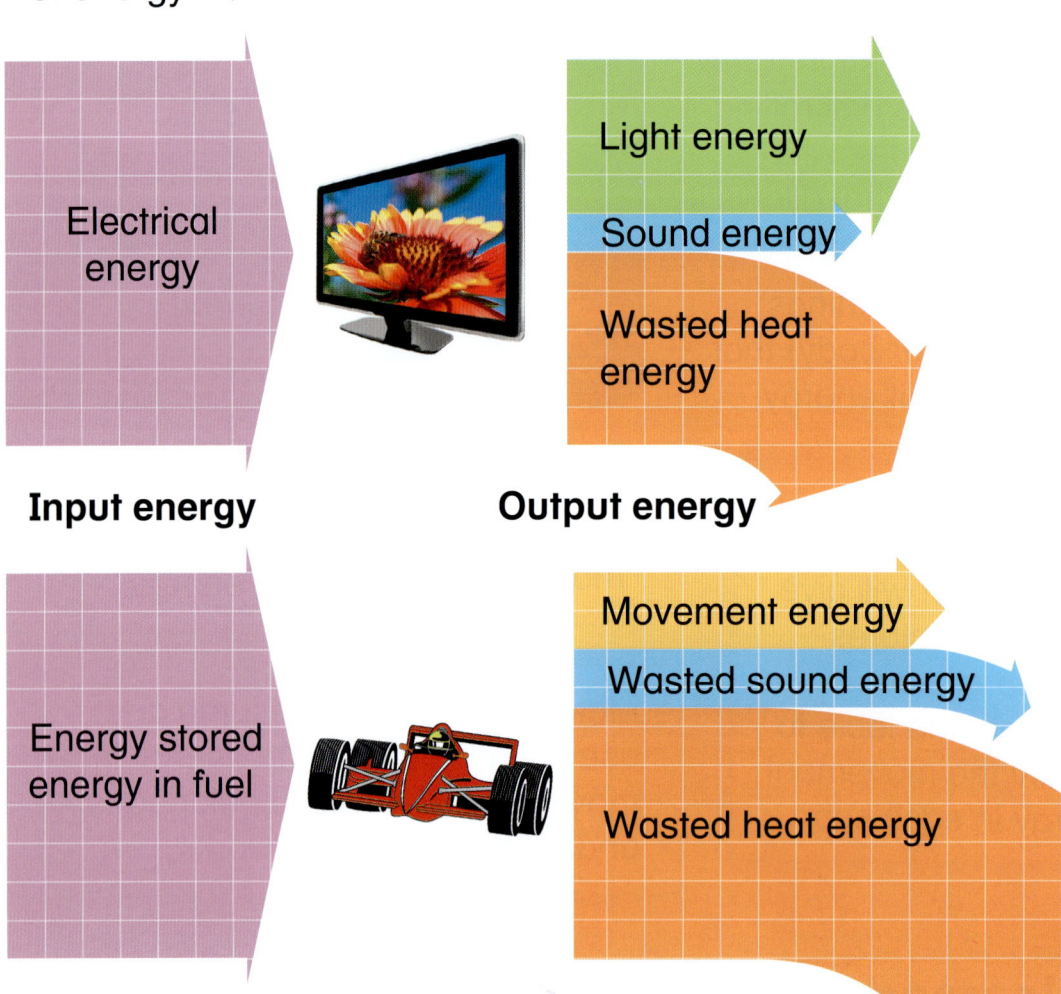

2 What forms of output energy are useful from a television? What form of output energy is wasted energy?

Topic **4** Energy and light

Some objects are designed to not produce a lot of wasted energy. These are called **energy savers** or energy-efficient items.

As well as giving off light, lamps can give off a lot of heat. For a lamp to be energy-efficient, it should not heat up too much when being used.

3 Using the table below, which lamp should be used? Explain your choice.

	Old fashioned lamp	Modern, compact lamp	Modern, energy saving, LED lamp
Energy used	40 watts	11 watts	7 watts
Lifespan	1 year	9 years	22 years

4 LED lamps are more expensive than old style lamps and compact lamps. Why do you think this is?

Activities

1 Describe the wasted output energy in the picture. Can you think of ways to reduce it?

2 Suggest some ideas for saving energy in your home and classroom.

3 Create an advertisement for an energy-saving appliance. Use the words below in your advertisement.

energy   efficient   money-saving

I have learned

- Energy is transferred from one object to another.
- Some energy is transferred to the surrounding environment during energy transfer.

4.5 How light travels

Key words
- light
- ray diagram
- shadow

Light sources provide the light we need to see properly. Light travels in a straight line from a light source to the objects we see. If there is no light source, it is dark and we cannot see.

1 Look at the photographs. What do you notice about the path a light beam takes? ▼

We can draw a **ray diagram** to show the direction the light is travelling in. The rays are drawn as straight lines with arrows to show the direction of the light. When a ray of light hits an object the light is blocked.

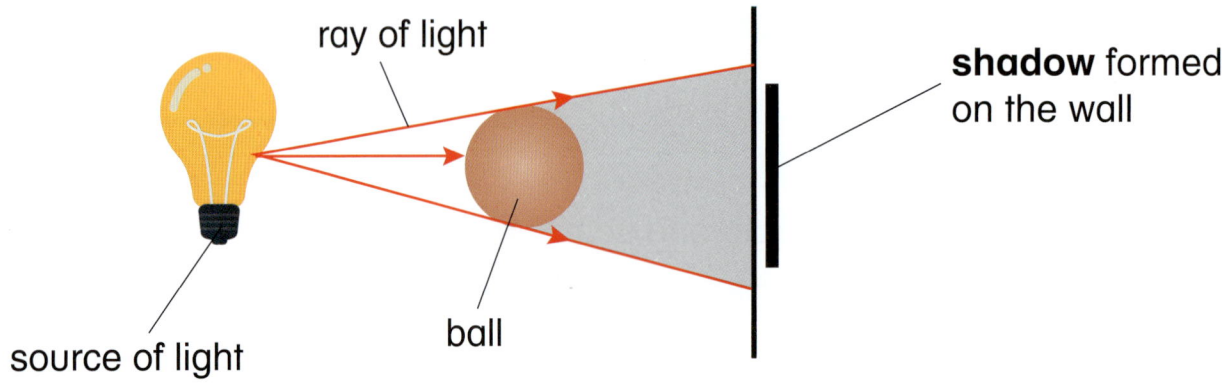

Topic **4**   **Energy and light**

2. What happens to a light beam when it hits a tree trunk? Draw a diagram to show what will be seen behind the tree. Add light rays to show how the light travels.

Shadows are formed when light travelling from a source is blocked. Some materials let some light through.

3. Predict what would happen if light rays could bend around the edges of objects, such as tree trunks.

Activities

1. Investigate whether light travels in a straight line. Use page 52 of your Workbook to record your investigation. Discuss your findings. Explain how this was a fair test.

2. Look at the diagrams on page 53 of your Workbook. Draw the rays of light from the light sources to the objects.

3. Conduct a different investigation which can test whether light travels in a straight line. Write your findings on page 54 of your Workbook.

I have learned

- Light travels in straight lines from a light source.
- Some materials block light because the light cannot travel through them; this causes shadows.

4.6 Seeing objects

Key words
- light source
- bounce
- reflect

Light travels from a **light source**. Objects that give out their own light are called light sources. The Sun, a flashlight, a lamp that is switched on, a flame from a candle and the stars in the sky are all light sources.

We can see objects because light from a light source shines on the objects, **bounces** (**reflects**) off them and enters our eyes. If there is no light, we cannot see at all.

▲ *This boy can see what is written in the book because light travels from the light source, is reflected off the book and then enters his eyes.*

Opaque materials block light that reaches our eyes. This can cause shadows. If a material is translucent it blocks some of the light that reaches our eyes. Light enters our eyes and falls on the inside parts of our eyes that are very sensitive to light. Translucent materials can therefore protect our eyes.

Topic 4 Energy and light

1. Work with a partner. Look carefully at your partner's eyes but don't touch them. Can you see the point where the light enters the eyes? It looks like a black spot, but it is really a hole.

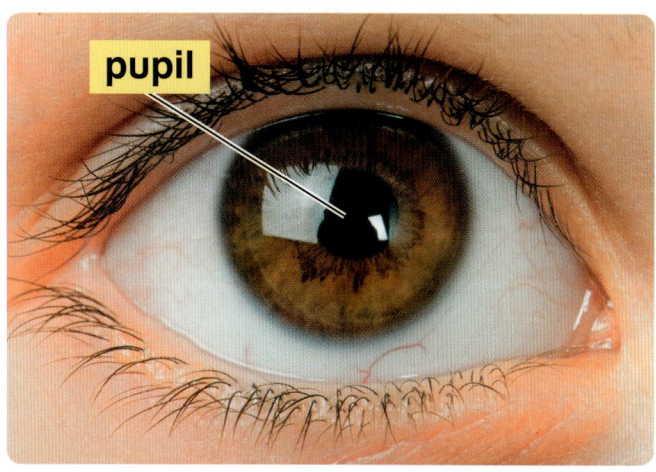

▶ The black spot (called the *pupil*) is where light enters the eye.

You already know that sunglasses are translucent and have a special coating on them which protects your eyes by blocking out harmful light from the Sun.

2. Explain how sunglasses protect your eyes.

Activities

1. Draw a picture of your partner's eye. Discuss why you think there is a hole in the middle of the eye.

2. Complete the diagrams and sentences on page 55 of your Workbook.

3. Draw two light sources on page 56 of your Workbook and explain why they are light sources.

I have learned

- We can see light sources because light from the source enters our eyes.

4.7 Reflecting light

We can see objects because light from a light source shines on the objects, **bounces** off them and then enters our eyes. When light bounces off a surface, we say that the light is **reflected** by the surface.

Key words
- bounce
- reflect

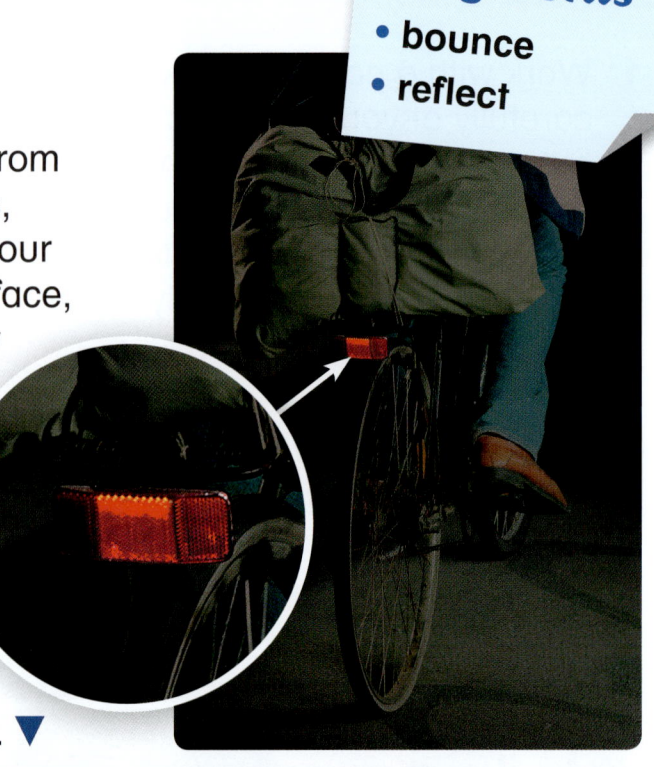

1. This bicycle reflector is not a light source, so how do we see it? Where does the light come from and how does it reach our eyes? ▶

2. Read this story. Discuss what you have learned from the story. ▼

In the stock room …

"I can't see anything."

"That's because we have no source of light in here."

Dalia switches on the flashlight.

"We can see the flashlight now."

"But we can see everything else too, even the things that aren't light sources."

Dalia has a bright idea.

"That must mean that all the other things reflect light because when the light source is off, we don't see them …"

"… and when the light source is on, we do see them."

Topic **4** Energy and light

The Sun is a light source and so we can see it clearly. The Moon is not a light source, but we can often see the Moon quite clearly in the sky. We can see the Moon when it reflects the light of the Sun.

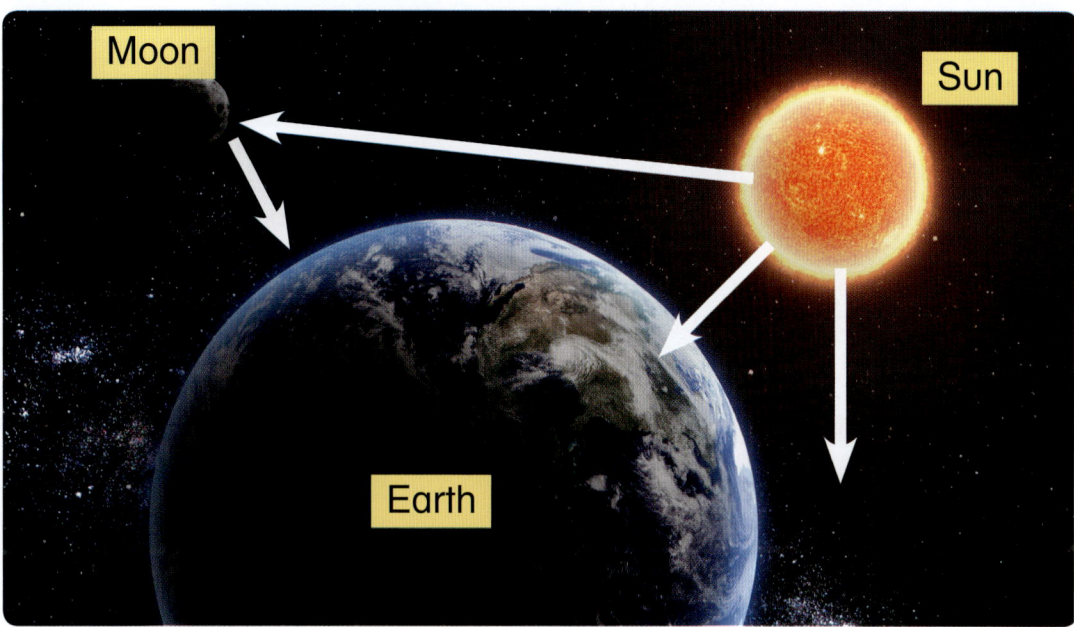

Activities

1 Draw a diagram on page 58 of your Workbook to show that you understand the way that light is reflected.

2 Create a comic strip story like the one on page 60 to help explain to younger children about light sources and reflected light.

3 Find out more about the way that the Moon reflects light from the Sun. Why can we sometimes only see part of the Moon? Can we ever see both the Sun and the Moon in the sky at the same time? Draw diagrams to explain your answers.

I have learned

- Light can be reflected from some surfaces.
- When light reflected from an object enters our eyes, we can see the object.

Looking back Topic 4

In this topic you have learned

- Energy is present in all matter. It is needed for any movement or action to happen.
- Energy cannot be made, lost, used up or destroyed, but it can be transferred. Heat, sound and light energy are examples of energy transfers.
- Not all energy transferred from one object to another comes out as useful energy. Some of the energy is wasted.
- Light travels in straight lines.
- We can see objects because light from a light source reflects off the objects and enters our eyes.

How well do you remember?

1. Explain the transfer of energy when a light is switched on in a house.
2. Identify the energy needed to bounce a ball.
3. Identify the light sources.

4. Draw a picture of a person with the Sun behind them. Include light rays in your picture. Label your picture and write two or three sentences to explain your drawing.
5. Draw a simple diagram to explain how we see things. Your diagram should include a light source and a simple drawing of the eye. Label your diagram and use arrows to show the direction of the light.

Topic 5 Electricity

In this topic you are going to learn more about electricity. You will work with simple circuits and design your own switch. You will look at the effect of using more components on lamps in a series circuit. You will identify good conductors and insulators of electricity.

Thinking and working scientifically

5.1 Wire investigation

Key words
- component
- pattern

You have already learned in Stage 2 that a circuit is a path through which electricity can move. Circuits need to be complete in order for lamps to light up. A complete circuit means that the **components** are all joined in the correct way and the circuit is closed.

Consider the following enquiry:

Does the length of the wire in a circuit change the brightness of a lamp?

Teri says if the wire connects correctly to form a complete circuit it does not matter how long it is. She claims the circuit below will work in the same way as one with a shorter wire.

1 What is Teri's prediction?

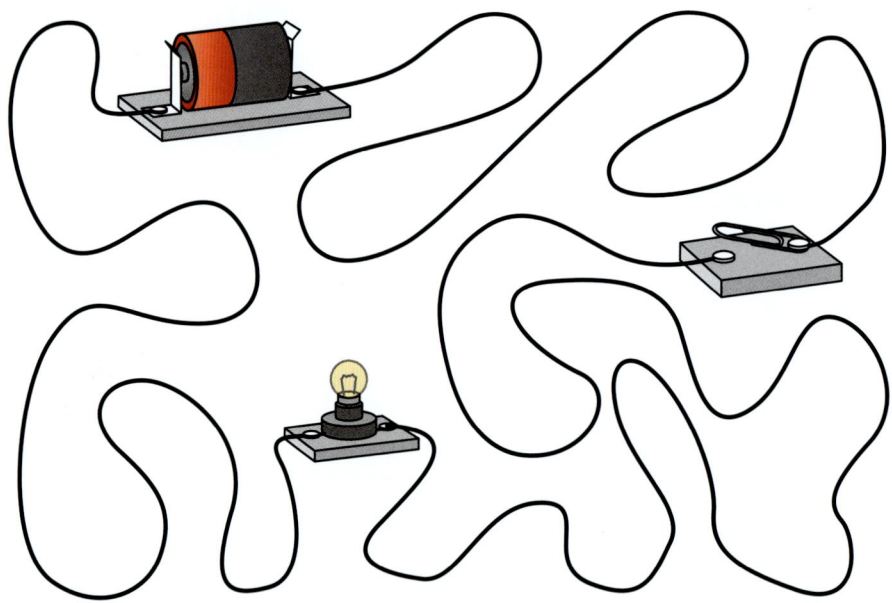

It is important to return to a prediction and see whether the results support or do not support it.

Having conducted the investigation, Teri realised that the longer the wire, the dimmer the lamp. With a short wire, the lamp shone more brightly.

2 Describe the **pattern** identified through this investigation.

Topic **5** Electricity

3 Do the results support Teri's prediction? Why or why not?

When conducting the investigation, Teri used different lengths of wire. She recorded her results in a table.

Length of wire	Very bright	Bright	Dim	Very dim
50 mm	X			
100 mm		X		
150 mm			X	
200 mm				X

4 Explain how Teri made sure this was a fair test.

Activities

1 Use components to build different circuits. Follow the instructions on page 60 of your Workbook.

2 Design a troubleshooting flow chart to show students what they should check if the lamp in a circuit does not light up. Use page 61 of your Workbook to record your work.

3 Does the length or path followed by the wires make a difference to a circuit? Read the statement and look at the diagram on page 62 of your Workbook and then do your own investigation.

I have learned

- Components must be connected correctly to form a circuit.

5.2 Why won't it work?

>Key words
>- circuit
>- break
>- switch

Electrical devices can only work if the electrical **circuit** to which they are attached is complete. If there is a **break** in the circuit, it is incomplete and the device will not work.

Consider the following problems.

Problem 1: Harry's flashlight won't light up.

Problem 2: Vanessa's reading lamp isn't working.

1 What could be causing each of these problems?

The devices in the pictures above are not working because there is a break in the circuit. There are a number of things that can cause a break in a circuit.

▶ **Switches** are used to break circuits on purpose. When the switch is off, the circuit is broken and the device will not work.

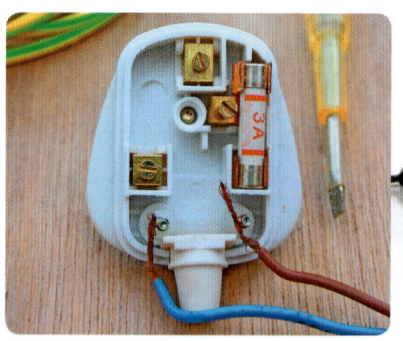

▲▶ Incomplete connections will also cause breaks in circuits. When a mains electrical device is unplugged, the connection to the electricity supply is broken and the device won't work. When wires or other components are loose or when there is something blocking the connection, the circuit is broken and the device won't work.

Topic **5** Electricity

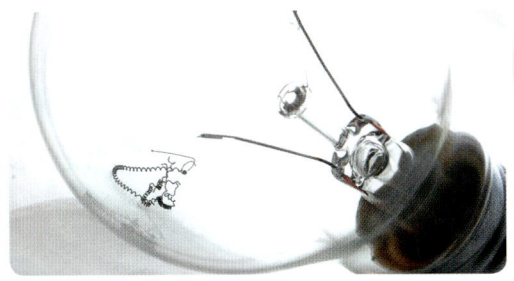

▲ Faulty components can cause breaks in circuits. Flat cells or batteries, broken or spent lamps, gaps between the connections in the switch and broken wires will all cause a break in the circuit and the device won't work.

▲ Special circuit breakers, such as fuses, are installed in homes and vehicles. If the electricity supply is too strong, or there is a problem with the supply, the fuse wire burns out and breaks the circuit by causing a gap that the electricity cannot flow through.

2. Many devices that you use at home rely on two switches to make a complete circuit. Explain how this works.

Activities

1. Write a set of instructions for Harry and Vanessa to explain what they should check to see why their electrical devices are not working. Refer back to your troubleshooting chart on Workbook page 61 if you need to.

2. Look at the electrical devices on the worksheet your teacher will give you and read the information next to each one. Suggest where the break could be in each circuit.

3. Naadira has a length of electrical cord. She thinks that some of the copper wires inside the cord may be broken. Explain how she could test this before she attaches the cord to her new reading lamp. Use Workbook page 63 to record your ideas and show your solutions.

I have learned

- An electrical device will not work if there is a break anywhere in the circuit.
- Switches are used on purpose to safely break and complete circuits.

5.3 Simple switches

An electrical **switch** allows us to control the flow of electricity in a circuit. When a switch is open there is a gap, which will **break** the circuit. When a switch is closed a **conductor** closes the gap, allowing the current to flow through the circuit again.

> **Key words**
> - switch
> - break
> - conductor

1 How does this switch work? ▼

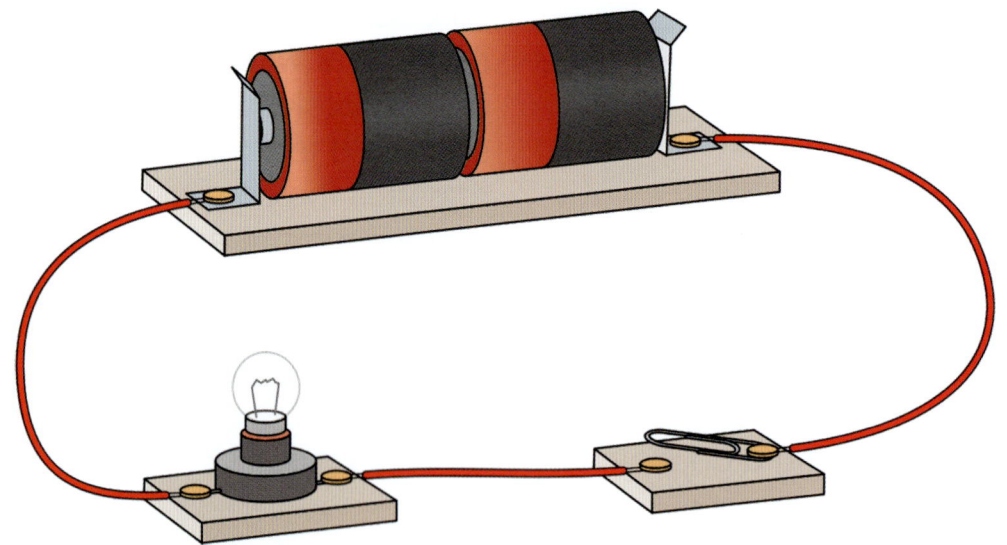

2 What do you think the switch in the circuit is made from? Why was this material chosen?

3 Would you expect the lamp to be lit in the circuit above? Explain your answer.

Topic **5** Electricity

Below are some pictures of different kinds of switches.

4 Where would you expect to find each of these kinds of switches?

a

b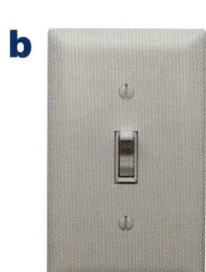

c

d

e

Activities

1. Compare the two switches on page 64 of your Workbook. Consider the materials used, how each switch works, and the advantages and disadvantages of each type. Use the table in your Workbook to summarise the information.

2. Can you design a better switch? Use page 65 of your Workbook to design a different switch. Present your ideas to the rest of the class and decide which switch would work best.

3. Many modern appliances can be switched on and off with a remote control. Find out more about how a remote control works and draw a diagram. Include labels and a few sentences to explain your findings.

I have learned

- A switch is used to open and close a circuit.
- You can use different materials to build a switch.

5.4 Series circuits

Look at this circuit:

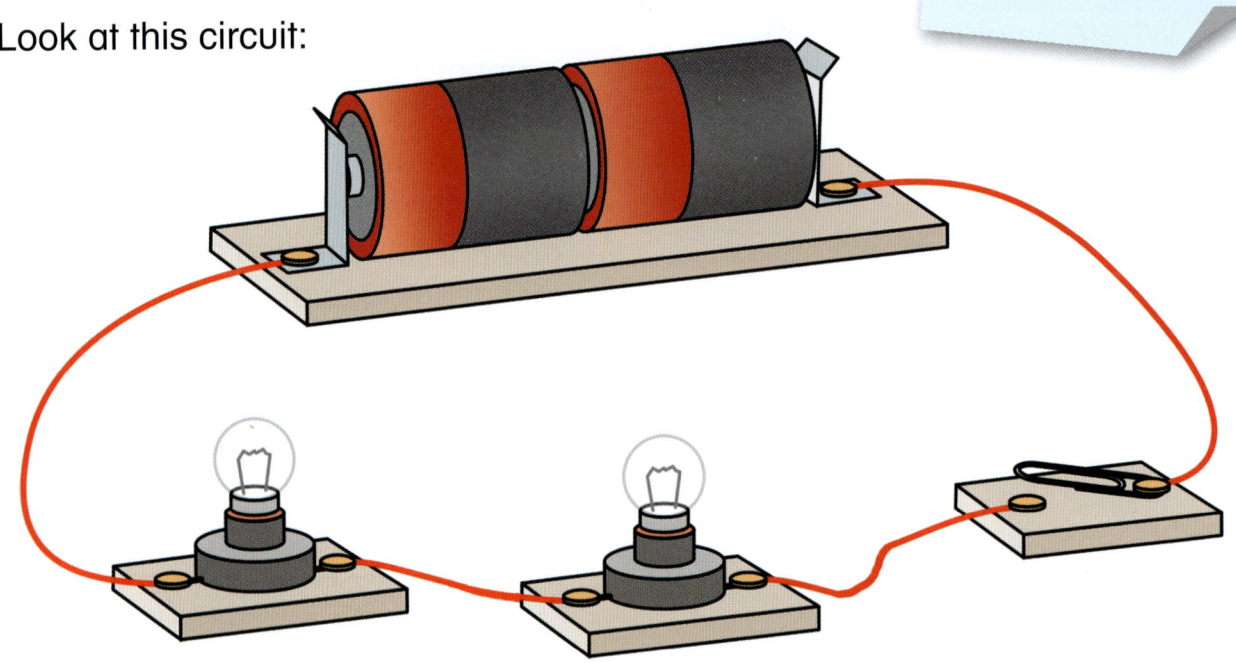

1. What **components** were used to build this circuit?
2. Describe how the components are joined in this circuit.

In this circuit the components are all joined in a loop, one after the other. A circuit like this is called a **series circuit**. In a series circuit, if you make a break anywhere along the loop, the current stops flowing – the lamps will not work if there is a break in the circuit.

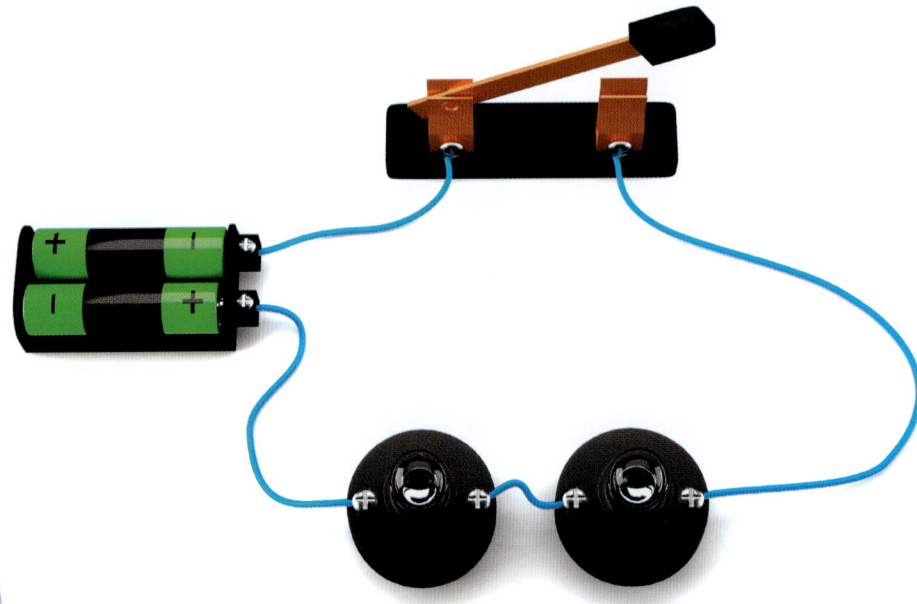

Topic **5** Electricity

Here are three series circuits with different components:

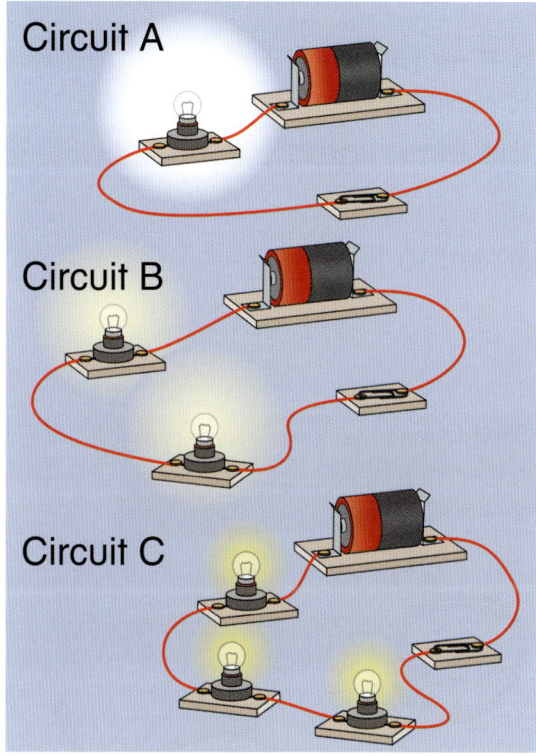

Circuit A

Circuit B

Circuit C

3 What do you think would happen if one of the lamps in circuit **B** was unscrewed? Why?

4 The lamp in circuit **A** seems to be brighter than the others. Is this likely to be the case? Explain why or why not.

5 How do you think adding a buzzer or a motor to each circuit would affect the brightness of the lamps?

6 How do you think adding a cell every time you added a lamp, buzzer or motor would affect the brightness of the lamp(s)? Why would this happen?

Activities

1 Carry out a test to observe what happens when you add lamps to a circuit without adding other components. Use Workbook page 66 to record your findings.

2 Build the circuits shown on Workbook page 67 and then add a cell to each. Record the effect this has on each circuit.

3 Does it matter whether you add a buzzer or a motor to a circuit with one lamp? Investigate how changing the type of component affects the lamp. Record your planning and results on Workbook page 68.

I have learned

- Changing the number or type of components affects the way a circuit works.
- Adding lamps without adding cells results in the lamps burning less brightly.

5.5 Conductors and insulators

Key words
- current
- conductor
- insulator

An electrical **current** can be thought of as a flow of electricity through the components in a path around a closed circuit. The current is flowing when the lamp lights up, the buzzer sounds or the motor runs.

Only certain materials allow an electrical current to flow through them. Look at these objects:

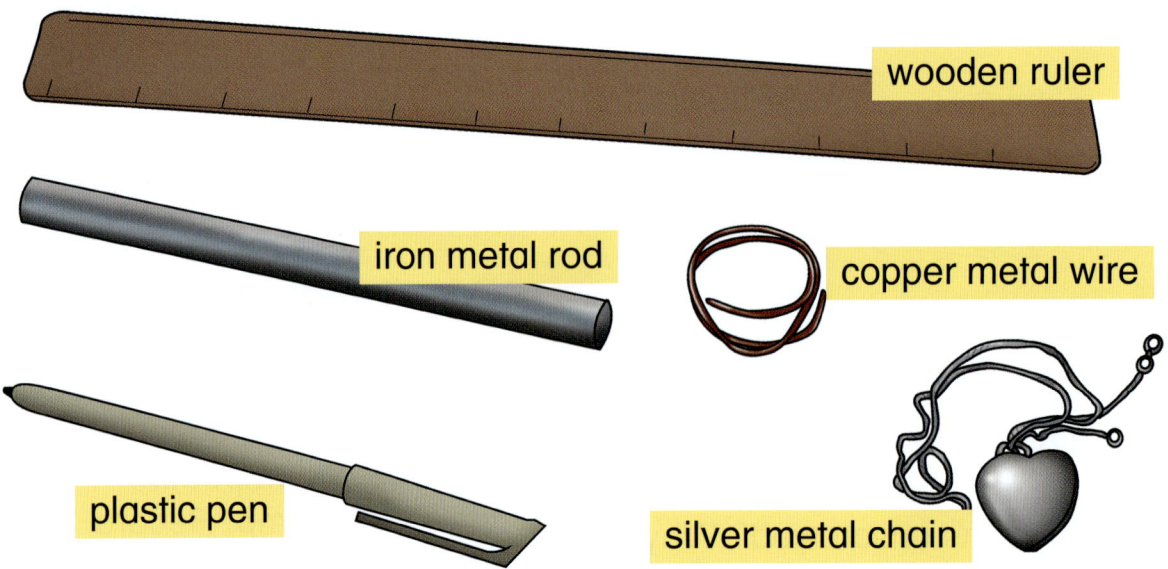

wooden ruler

iron metal rod

copper metal wire

plastic pen

silver metal chain

1 What could you do to find out whether electricity will flow through these objects? Share your ideas with the class.

A material that allows a current to pass through it is called a **conductor**. You already know that wire is able to conduct a current. You can test other materials by connecting them to a simple circuit with one cell and a lamp to check whether the lamp lights up. If the lamp lights up, the material conducts electricity; if not, the material does not conduct electricity.

The best conductors of electricity are metals. Water is also a conductor of electricity. This is why you should never use electrical appliances in the bath or while standing on a wet floor.

A material that does not allow a current to flow through it is called an **insulator**. Plastic and rubber are good examples of insulators.

Topic **5** Electricity

Insulators are used to make sure that electrical current does not flow to a place where it is not wanted or where it could cause injury or damage.

2 Why do electricians working on electric cables wear thick rubber gloves and safety shoes with thick rubber soles?

3 Why should you get out of the water if you are swimming and there is a thunderstorm?

The electricity we use in our homes flows through thick metal cables held up by strong poles. This current is strong enough to kill you. To make the power lines safe, special insulators made from glass or ceramic are placed between the power lines and the poles, to stop the metal poles from conducting the current.

Activities

1 Sketch or photograph the electricity poles in your area. Label your diagram to show the conductors and the insulators.

2 A lamp contains materials that are good conductors and others that are not. Label the diagram on page 69 of your Workbook to show where each of these is found in a lamp.

3 Why does a lamp need to conduct current in some places but not in others? Record your answers on page 69 of your Workbook.

I have learned

- Materials that allow electricity to flow through them are called conductors. Metals are good conductors.
- Materials that do not conduct an electrical current are called insulators. Rubber and plastic are good insulators.

Science in context

Key word
- lightning

5.6 Lightning conductors

Lightning is a very strong flash of electricity. Lightning strikes can start fires and injure people. Some places are struck by lightning more often than others, for example, tall buildings. To lessen or stop lightning from damaging a building, a lightning conductor is used.

Look at the diagram below, which shows how a lightning conductor works.

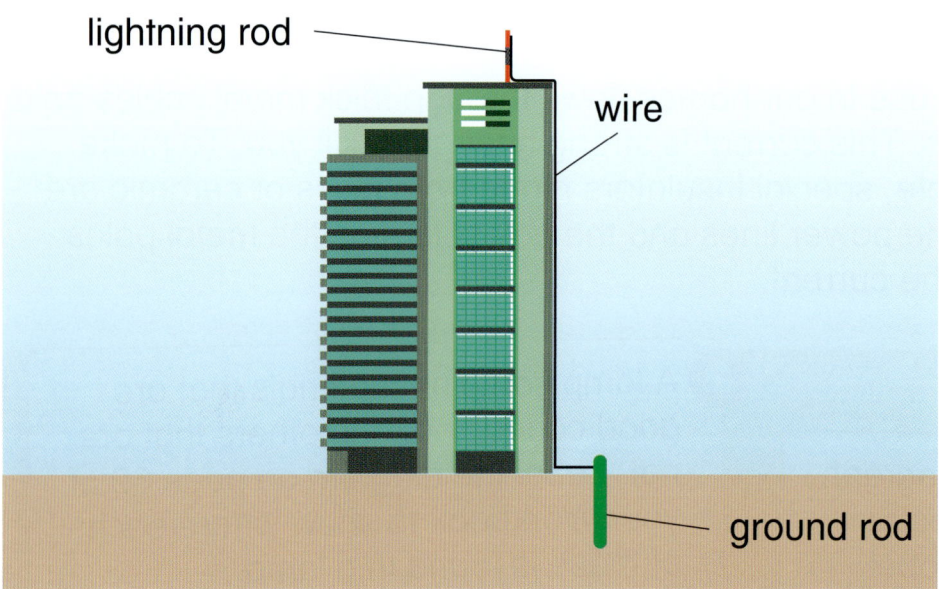

1 Why are lightning conductors positioned on the tops of buildings?

Before the development of modern building techniques, buildings could not be built as tall as they can be now. Scientists had to find a way of preventing buildings from being struck by lightning. Today, architects and builders make sure that there are lightning conductors to protect tall buildings.

2 What is the purpose of the ground rod?

If lightning hits a building which has a lightning conductor, it will usually hit the highest point on the building. The electric current will strike the conductor and travel along the wire to an area under the ground. This is the earth connection which protects the people and building from harm and damage.

74

Topic **5** Electricity

Activities

1 Using a diagram, explain how a lightning conductor works.

2 Design your own lightning conductor. Add labels explaining what each part is and what material the conductor is made from.

3 Research Benjamin Franklin and the history of lightning conductors. Prepare a short presentation to give to your class.

3 Why do you think it is important that a building with a thatched roof has a lightning conductor?

Lightning conductors come in many different shapes, with the conductive material being the only thing in common. Examples of materials used are copper and aluminium.

4 Why are lightning conductors not made from plastic?

I have learned

- Lightning conductors help protect buildings from getting damaged and people from getting electrocuted.
- Electrical conductors are usually made from copper or aluminium.

Looking back Topic 5

In this topic you have learned

- Circuits are built with wires, cells, lamps and switches.
- A circuit must be complete and closed for the electricity to be able to flow around it.
- A simple switch is used to control the flow of the current through a circuit.
- The more lamps or other components added to a series circuit, the dimmer the lamps will become.
- Metals are good conductors of electricity.
- Plastic and rubber are good insulators of electricity.

How well do you remember?

1 Explain how you can tell whether an electrical current is flowing in a circuit.

2 Nadia's electric keyboard won't work when she switches it on. Suggest two things she should check.

Look at the circuits on the right.

3 In which circuit will the lamps shine the brightest? Why?

4 In which circuit will the lamps be the dimmest? Why?

5 Explain why adding a buzzer to circuit **A** would affect the brightness of the lamp.

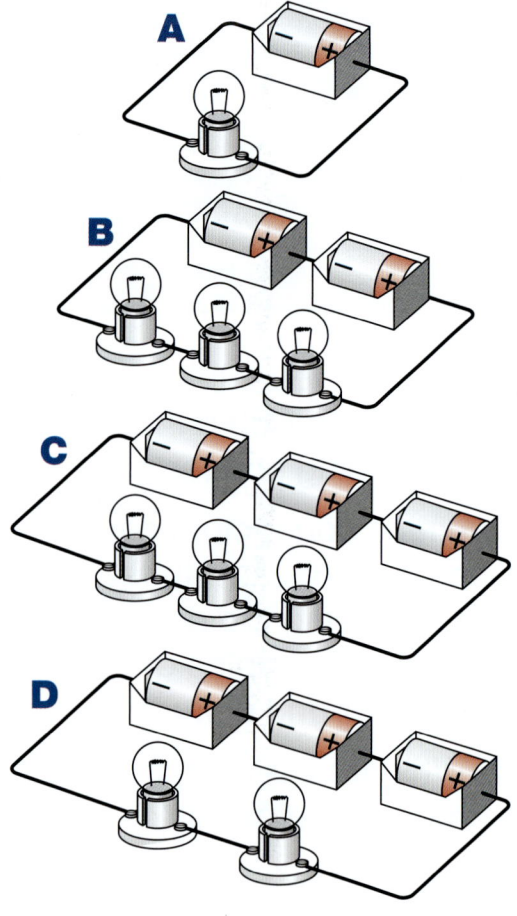

Topic 6 Planet Earth

In this topic you will learn more about the Earth. You will describe the internal structure of the Earth, and you will find out what is happening below the surface of the Earth to cause earthquakes and volcanoes.

6.1 Earth's structure

Key words
- crust
- mantle
- core

Planet Earth is shaped roughly like a sphere and is made up of different layers. The diagram below is a representation of what you would see if you were able to cut a slice through the Earth.

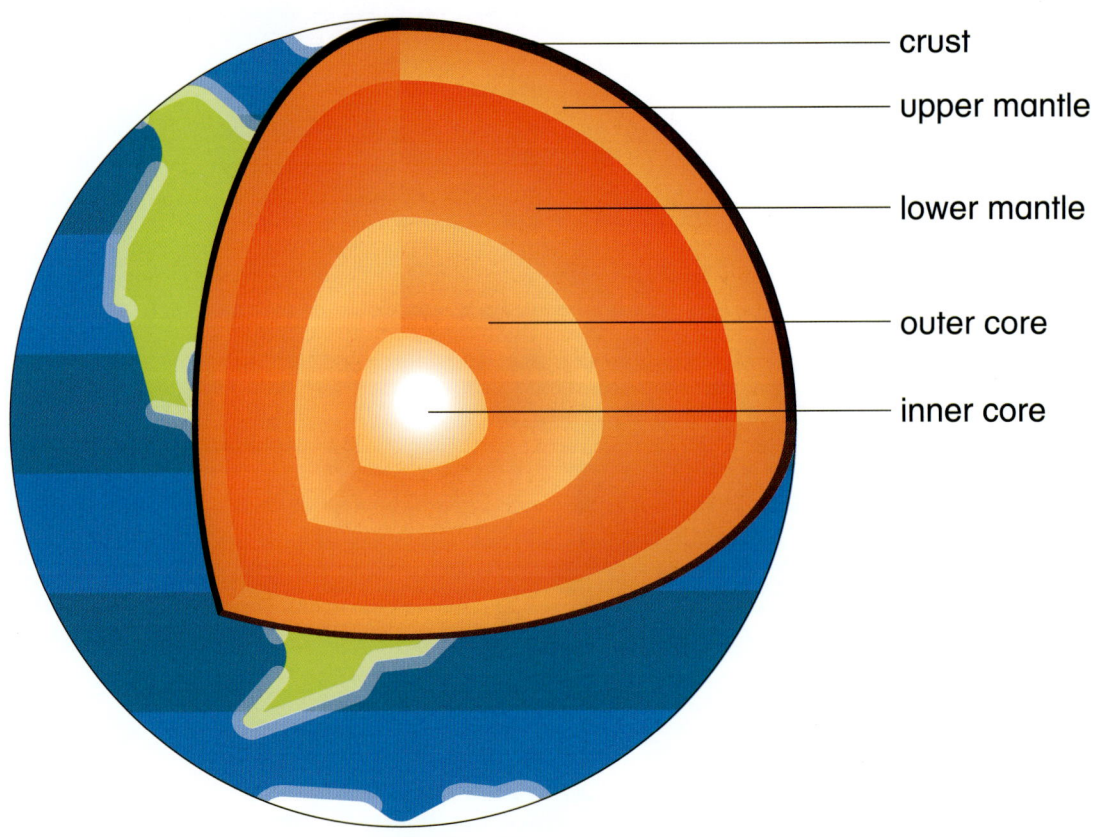

- crust
- upper mantle
- lower mantle
- outer core
- inner core

1. How many layers does the Earth have? On which layer do we live?
2. Why do you think the scientists labelled them as different layers?

The thinnest layer of the Earth's surface is the **crust**, which is where we live. This layer is made of rock and can be divided into continental crust (land) and oceanic crust (sea). They are different because the main type of rock on these two crusts is different.

Topic 6 Planet Earth

The **mantle** is divided into the upper and lower mantles. The upper mantle is just below the crust and is made up of hot rock. The lower mantle is mainly made up of melted rock.

The centre of the Earth is called the **core**. This is the hottest layer of the Earth. It is divided into the outer and inner core. The outer core is made of liquid metal and the inner core is made of solid metal. It is the metals in the inner core which create the magnetic field on Earth.

3 Why do you think the inner core is solid metal, even though it is extremely hot?

Activities

1 Use books or the internet to find out the temperatures of the different layers of the Earth. Record these on page 70 of your Workbook.

2 Complete a diagram showing the different layers of the Earth's surface.

3 Use modelling clay to create a model of the structure of the Earth. Add labels for the different layers. Do some research and write one interesting fact about each layer.

I have learned

- The structure of the Earth is made up of different layers.
- We live on the surface of the Earth; this is called the crust and is the thinnest layer.
- The mantle is divided into the upper and lower mantle.
- The Earth's inner layer is known as the core and is divided into the outer and inner core.

6.2 Volcanoes

Key words
- volcano
- molten rock
- lava
- tectonic plates

There are more than 1500 active **volcanoes** on the Earth. Some of these volcanoes are in the oceans. The largest active volcano is in Hawaii.

A volcano is a mountain that opens downwards to a pool of **molten rock** below the Earth's crust. When the pressure builds up under the surface of the Earth, the mountain will erupt. Gas and rock shoot up out of an opening at the top of the mountain and **lava** flows down the slopes of the mountain.

A cross-section through a volcano

Labels: dust, ash, steam and gas; crater; main vent; side vent; layers of ash and cooled lava; Earth's crust; magma chamber; lava

1 Use the diagram of a volcanic eruption to explain what happens when the lava pours out.

The Earth's crust is like a jigsaw puzzle, with the huge slabs of crust fitting together. These huge slabs are known as **tectonic plates**. The plates are all moving very, very slowly across the surface of the Earth.

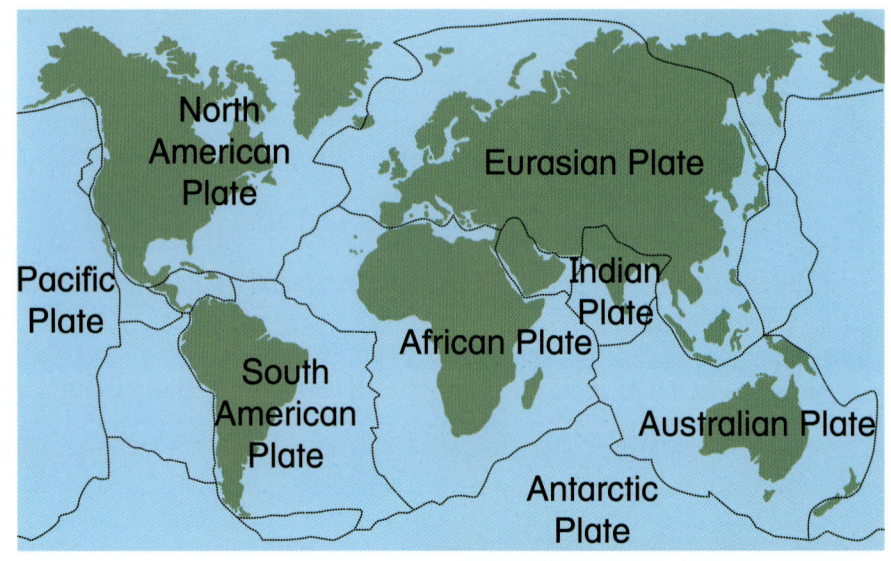

The Earth's tectonic plates

Topic **6** Planet Earth

Where the edges of the tectonic plates meet each other and collide the land is pushed upwards to create mountains. Where the edges are pulling apart from one another they create deep valleys. It is the movement of these plates which can cause a volcanic eruption.

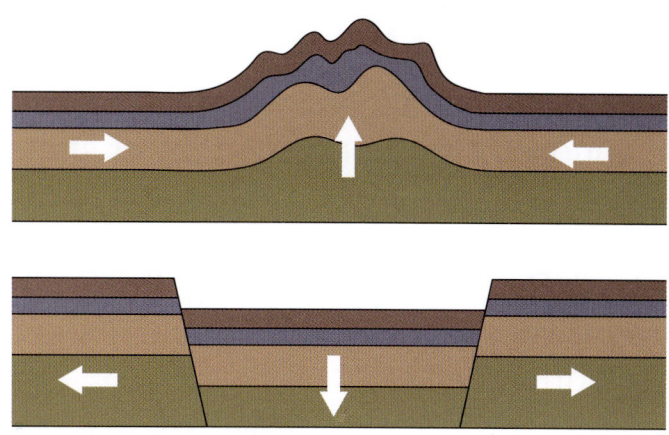

2 Volcanoes occur on mountains on the surface of the Earth. Where do they occur in the oceans?

3 The ash and lava from a volcano can wipe out everything in the local area. How should people prepare for these eruptions if they live near an active volcano?

Activities

1 Paint and label a cross-section through a volcano. (Make sure you close your Student's Book so you are doing it from memory!)

2 Research where active volcanoes can be found. Complete the map on page 72 of your Workbook to show where three of them are located.

3 Do some research on the 'Ring of Fire' in the Pacific Ocean. What is it and where can it be found? Use Workbook page 73.

I have learned

- Volcanoes occur due to the movement of the tectonic plates on the Earth's surface.
- When pressure builds, molten rock erupts into the air, with ash and lava pouring out of the top of the mountain.

6.3 Earthquakes

Key words
- stress
- fault line
- tsunami
- aftershock

An earthquake happens when the tectonic plates on the surface of the Earth squeeze or stretch, causing the rocks at the surface to lift or move apart. This movement is what makes the Earth's surface shake and roll. By shaking and rolling, the Earth gets rid of **stress**. These stresses are found at the **fault lines** in the Earth's crust.

1 What damage can be caused by an earthquake?

2 Why are scientists unable to always warn people of earthquakes?

An earthquake can occur on land or in the ocean. When an earthquake occurs in the ocean, it can cause a **tsunami**. This can lead to a huge wave destroying the land near to it.

There are more than a million earthquakes every year, but they usually only last for less than a minute, and some are so small you cannot feel them.

Topic **6** Planet Earth

The strength of an earthquake is measured using the Richter Scale.

Minor Usually cannot be felt

Light Can be felt but causes little damage

Moderate Some slight damage to buildings

Strong Can cause damage to buildings

Major Causes serious damage

Great Can destroy whole areas

2   3   4   5   6   7   8

Aftershocks can happen after a large earthquake. These are like smaller earthquakes and affect the same place where the earthquake took place. Depending on how big the original earthquake was, aftershocks can continue for days afterwards.

3 Why do earthquakes not occur everywhere on Earth?

Activities

1 Can you identify any similarities between an earthquake and a volcano? Compare them in the table on page 74 of your Workbook.

2 Design a poster for a classroom where earthquakes are common. Use drawings and instructions of what the children need to do if there is an earthquake.

3 Do some research to find out what a seismograph is. Then build your own seismograph.

I have learned

- An earthquake occurs because of the movement of the tectonic plates.

Science in context

Key words
- sensor

6.4 Earthquake alerts

Scientists are developing a machine which can warn people of an earthquake before it happens. This is an important development, as many people die from being trapped under buildings which collapse during earthquakes.

1 Why is the development of an earthquake warning machine important?

It is still not yet possible to predict when an earthquake is going to happen. However, by using **sensors** which have been put in the ground, scientists know when an earthquake has started and that shaking is going to begin.

The sensors send a signal to an earthquake alert centre. The centre then sends information out to the people in the area through their cellphones and computers to warn them.

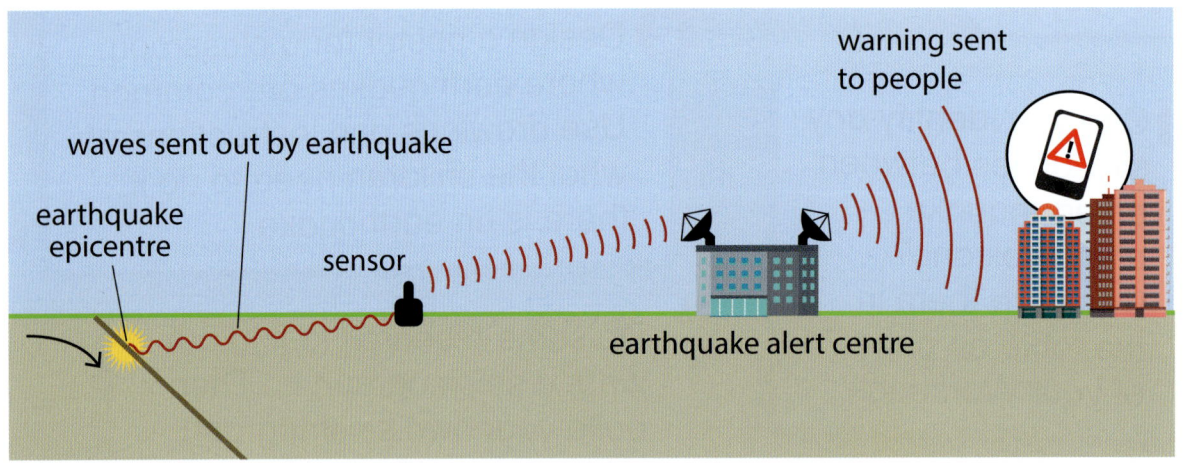

2 How can earthquake alerts help people prepare?

Tsunami warnings work in a similar way. When an earthquake under the ocean is detected by scientists, a warning is sent to people living in coastal areas that might be affected by the huge wave.

Topic 6 Planet Earth

1 *Two plates collide and an earthquake happens under the ocean. The water is pushed upwards very quickly. On the shore the water is sucked back out to sea.*

Earthquake

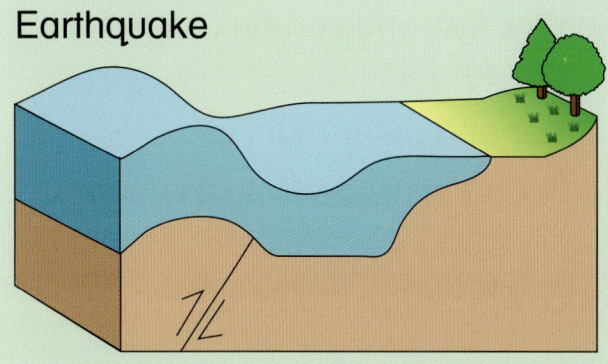

2 *Gravity pulls the water downwards and a tall tsunami wave is formed, which rushes towards the shore.*

Tsunami wave surge

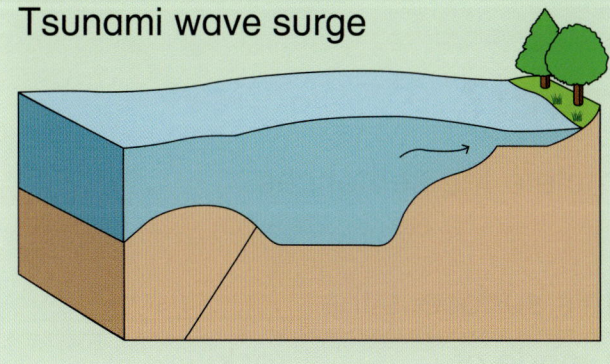

3 How do you think these alert systems save lives?

Activities

1 Research your local area. Do you live in an area that has earthquakes? Are there any alert systems in place?

2 Design a house and conduct the earthquake investigation on page 75 of your Workbook.

3 Improve on your house design from Activity 2 and test it on 'jelly Earth'. What do architects need to keep in mind when designing houses in an earthquake area? Record your results on Workbook page 76.

I have learned

- Scientists use sensors in the ground to detect earthquakes starting.
- An earthquake alert centre sends out warnings about earthquakes and tsunamis to people living in the area.

85

Looking back Topic 6

In this topic you have learned

- The structure of the Earth is divided up into the core, mantle and crust.
- The crust is the surface on which we live and can be divided into the continental and oceanic crusts.
- The mantel is divided into the upper and outer mantles.
- The core is the centre of the Earth and is divided into the outer and inner cores.
- Volcanoes and earthquakes occur where the plates of the Earth's crust meet.
- A volcano erupts when pressure builds up in the crust and hot magma, rock and ash escape.
- Volcanoes occur in the ocean and on land.
- Earthquakes are the shaking of the Earth's surface due to pressure being released.

How well do you remember?

1 Choose the correct words to complete the sentences.

   shake   crust   solid

   a The thinnest layer of the Earth's surface is the _____.
   b The inner core of the Earth is made of _____ rock.
   c An earthquake causes the ground to _____.

2 Explain what happens under the ocean to cause a tsunami.

3 Which of these buildings would be safest in an earthquake area?

A
B
C

Topic 7 Earth and beyond

In this topic you will learn more about the way the Earth spins, and how we have day and night. You will learn the names of the planets in the Solar System and you will focus on the importance of the Sun. You will learn that planetary systems can contain stars, planets, asteroids and comets.

7.1 Day and night

Key words
- sphere
- horizon
- model

Scientists now know that the Moon moves around the Earth and that the Earth moves around the Sun, but for a long time people thought that the Sun moved around the Earth. This is because it appeared to travel across the sky from the east to the west every day.

▲ In this picture, a photograph of the Sun was taken every 30 minutes and the images were put together to show the way in which the Sun rises and appears to move across the sky from east to west.

Today, we still say that the Sun 'rises' at a time that we call sunrise. And we say that the Sun 'sets' at a time that we call sunset, even though we know that it is really the Earth that moves and not the Sun. The Sun rises in the east and sets in the west.

1 Why does the Sun appear to move across the sky?

The Earth is shaped like a **sphere**. It moves around the Sun once a year and it also spins as it moves, in an *anti-clockwise* direction (from west to east).

Topic **7** **Earth and beyond**

As the Earth spins around and parts of it come to face the Sun, the Sun seems to rise up in the sky. This creates long shadows on the surface of the Earth. As the Earth spins away from the Sun, the Sun seems to sink down below the **horizon**, once again creating long shadows, until the Sun disappears.

2 What does a sphere look like?

3 What direction would the Earth be spinning in if it moved from east to west?

We can **model** the Earth spinning to show the reason why we get night and day. When it is daytime where we live, it is night-time on the opposite side of the Earth.

4 Where does the Sun go at night?

5 If it is daytime where you live, is your part of the Earth facing the Sun or facing away from the Sun?

▲ This photograph shows the Earth as seen from space. The countries that are facing the Sun have daylight. The other countries are in darkness, so there it is night-time.

Activities

1 Follow the instructions on PCM ES8 to build a papier mâché model of the Earth.

2 In your groups, use your model to demonstrate that the Earth is spinning and the Sun stays still, so that when one part of the Earth is facing the Sun it is daytime and on the other side it is night-time.

3 Use your model to explain why shadows change on Earth at different times of the day.

I have learned

- The Earth moves around the Sun.
- The Sun rises in the east (sunrise) and sets in the west (sunset).
- We can use a model to show that the way the Earth spins causes day and night on Earth.

7.2 The Earth rotates on its axis

Key words
- rotate
- axis
- angle

Think about the way in which you modelled the movement of the Earth in Unit 7.1. What did you do to make the Earth spin or turn around? You probably put a stick through the middle of your model of the Earth.

The globe in the picture has a stick through its middle too. But of course there is no stick through the centre of the real Earth!

The Earth spins or **rotates** on its **axis**. This is an imaginary line that runs from the North Pole through the centre of the Earth to the South Pole.

▶ A globe is a three-dimensional model of the Earth.

It takes the Earth 24 hours (one day) to complete one rotation; in other words, to turn around completely on its axis once.

1 How many complete turns on its axis will the Earth make in one week?

2 What would happen if the Earth did not rotate on its axis?

If you modelled the Earth carefully, you will have placed your stick at an angle because the axis of the Earth is tilted at an angle. Look at the picture of the globe and the diagram below. Can you see which way the Earth is tilted?

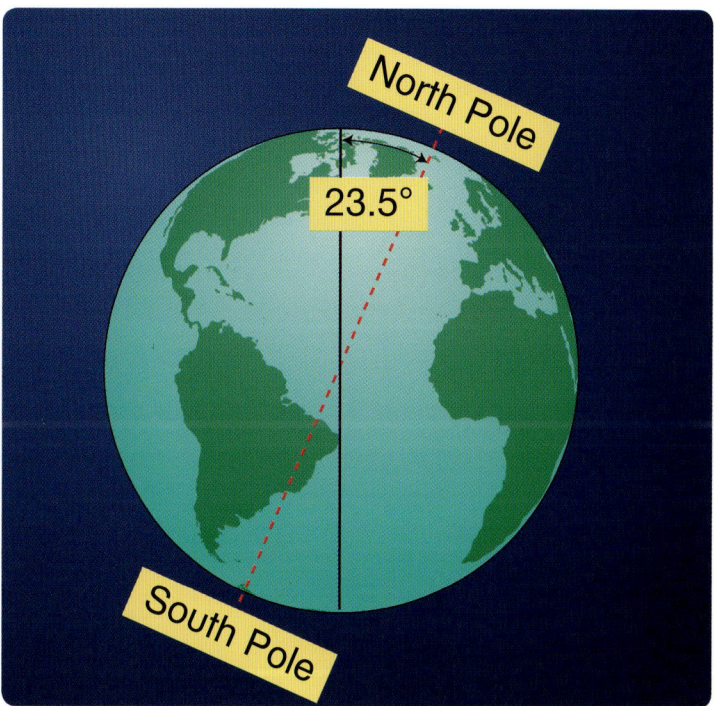

▲ *The Earth is tilted at an angle.*

3 What angle is the Earth tilted at?

4 What would happen if the Earth took 12 or 36 hours to turn on its axis?

Activities

1 Read the statements on page 77 of your Workbook and say if they are true or false.

2 Answer the questions in your Workbook on page 78.

3 Work in groups. You know that the Earth moves around the Sun. Think about the way the Earth is tilted on its axis. What effect do you think this angle has on life on Earth? What would happen if the Earth were not tilted at an angle? Use the model you have made or a globe to help you.

I have learned

- The Earth turns around, or rotates, on its axis.
- The Earth's axis is tilted at an angle.
- The Earth takes 24 hours to make one complete rotation.
- We have day and night on Earth because of the Earth's rotation on its own axis.

7.3 The Solar System

The Earth is not the only **planet** to **orbit** the Sun. There are other planets and there are also **moons**, which orbit around the planets. The Sun, the planets and their moons are all part of the **Solar System**.

Key words
- planet
- orbit
- moon
- Solar System
- gravity

The Solar System is made up of eight planets, with the Sun at the centre. There are four inner planets, which are the planets closest to the Sun. These are Mercury, Venus, Earth and Mars. The four outer planets, which are furthest from the Sun, are Jupiter, Saturn, Uranus and Neptune.

1. Make up a rhyme to help you to remember the names and order of the planets.

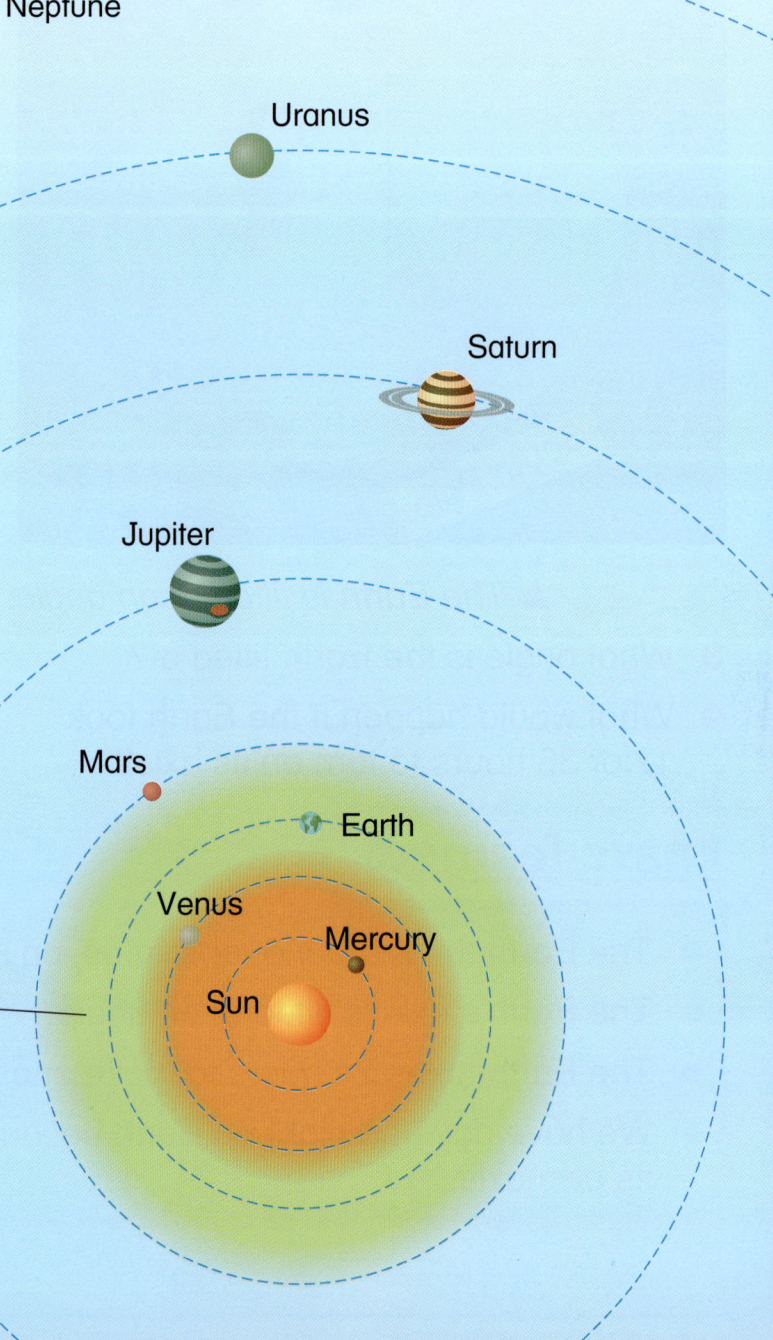

The green colour shows the region of the Solar System where a planet can have liquid water.

life. It also has liquid water on its surface, and life needs water. Another important thing about Earth is that it has an atmosphere. This is the mixture of gases surrounding Earth. Without these gases, the heat from the Sun would be lost from Earth and it would be much, much colder at night-time.

2 Explain the factors that make Earth a suitable planet for plants and animals to live.

The Sun is not a planet, it is a star. It is a hot ball of glowing gas at the centre of the Solar System. The pull of the Sun's **gravity** keeps all the planets in orbit around it. The Earth and all the other seven planets move in an orbit around the Sun. An orbit is a path that an object takes around something else.

3 Why do the planets not crash into each other?

1 Create a classroom display of the Solar System to show the Sun and the eight planets which orbit it. Most of the planets have their own moons. Include any that you find out about on your display.

2 Do some research to find out about one of the planets in the Solar System. Create a poster and present your findings to the class.

3 Research some facts about the sizes of the planets in the Solar System. Think of a way to display this information in a bar chart or table.

I have learned

- The Earth is the only planet in the Solar System which has perfect conditions for life.
- There are eight planets in the Solar System. They all orbit the Sun.
- The Sun is at the centre of the Solar System.

Topic 7 Earth and beyond

one of bill...
systems in the **Universe**. The Solar System is in the Milky Way. This is the name given to the **galaxy** we live in. Each galaxy contains millions and billions of stars, just like our Sun. Galaxies are huge. Our galaxy is one of billions that make up the Universe.

A planetary system has a star at the centre with orbiting planets, just like our Sun. But planetary systems also contain other objects such as moons, **asteroids**, meteoroids and **comets**.

Earth, Mars, Jupiter, Saturn and Neptune all have orbiting moons.

An asteroid is an uneven chunk of rock and metal. They vary in size from a few centimetres to hundreds of kilometres. In the Solar System, asteroids orbit the Sun in a 'belt' between the orbits of Mars and Jupiter.

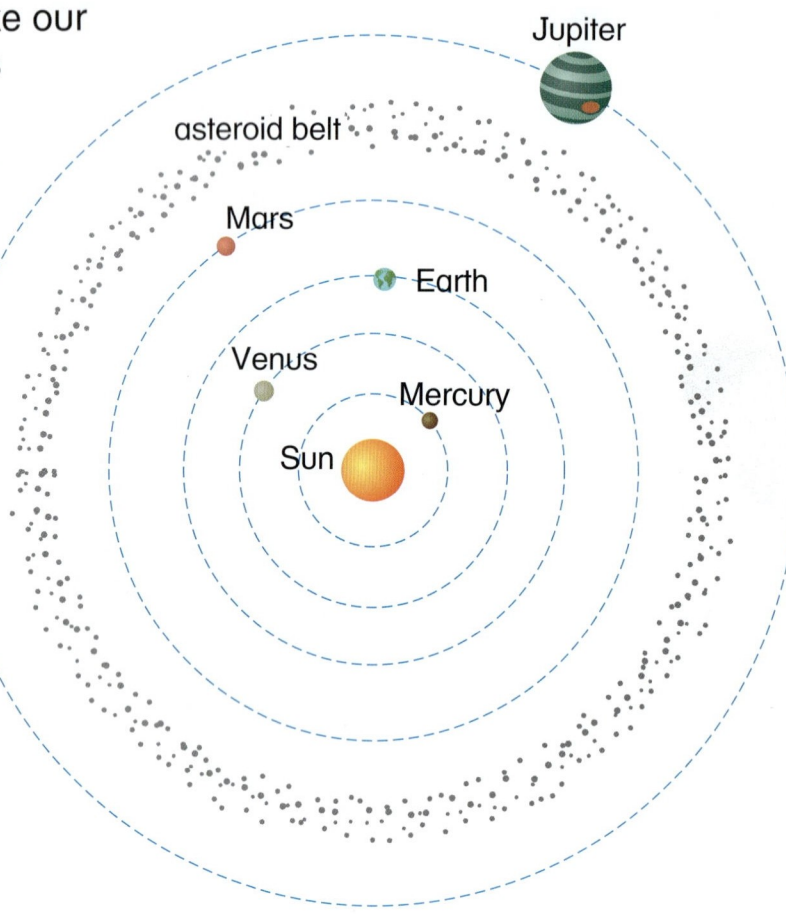

Acknowledgements

The publishers wish to thank the following for permission to reproduce photographs. Every effort has been made to trace copyright holders and to obtain their permission for the use of copyright materials. The publishers will gladly receive any information enabling them to rectify any error or omission at the first opportunity.

p1 Anna Omelch fenko/Shutterstock, p2tr Gelpi JM/Shutterstock, p2bl In Green/Shutterstock, p2br Aletia/Shutterstock, p3t dampphoto/Shutterstock, p3b Nicholas Piccillo/Shutterstock, p4tr Image Point Fr/Shutterstock, p4cr AN NGUYEN/Shutterstock, p4bl Jaimie Duplass/Shutterstock, p4bc Kozini/Shutterstock, p4br Jaimie Duplass/Shutterstock, p6t AJCespedes/Shutterstock, p6tc MARGRIT HIRSCH/Shutterstock, p6bc Elena Masiutkina/Shutterstock, p6bc Fotokostic/Shutterstock, p7 Photoagriculture/Shutterstock, p8tr Georgios Kollidas/Shutterstock, p8tl 2d Alan King/Alamy Stock Photo, p8br Everett Historical/Shutterstock, p8bl Primeiya/Shutterstock, p9t fotohay/Shutterstock, p9b didesign021/Shutterstock, p12l zlikovec/Shutterstock, p12r Janelle Lugge/Shutterstock, p13l travelfoto/Shutterstock, p13r Vaganundo_Che/Shutterstock, p14l 3DMI/Shutterstock, p14c effective stock photos/Shutterstock, p14r Richard G Smith/Shutterstock, p15t Johannes Dag Mayer/Shutterstock, p15c paulrommer/Shutterstock, p15b photonewman/Shutterstock, p16l Peter Gudella/Shutterstock, p16r Roblan/Shutterstock, p17b bluefish_ds/Shutterstock, p17c Ekaterina V. Borisova/Shutterstock, p17t Anan Kaewkhammul/Shutterstock, p19 Vibrant Image Studio/Shutterstock, p20l Wellcome Library, London. Wellcome Images, p23t design36/Shutterstock, p23b YanLev/Shutterstock, p27 Tiut Vladut/Shutterstock, p28tl photowind/Shutterstock, p27cl photowind/Shutterstock, p27tr Iakov Filimonov/Shutterstock, p27cr srdjan draskovic/Shutterstock, p27br scubaluna/Shutterstock, p29t wacpan/Shutterstock, p29c Aleksandar Dickov/Shutterstock, p29b Cloudpost/Shutterstock, p31 beboy/Shutterstock, p32 jarabee123/Shutterstock, p34l Sever180/Shutterstock, p34c Khvost/Shutterstock, p34r nexus 7/Shutterstock, p39t Brian Mueller/Shutterstock, p39bl sjeacle/Shutterstock, p39br RF97/Shutterstock, p40 Penny Hillcrest/Shutterstock, p40t assistant/Shutterstock, p40c Andrey N Bannov/Shutterstock, p40b Sashkin/Shutterstock, p42tl Lukas LK-Arts/Shutterstock, p42tr Valentyn Volkov/Shutterstock, p42bl Martina_L/Shutterstock, p42br suradech sribuanoy/Shutterstock, p43tl Valentin Kundeus/Shutterstock, p43tr Andrey Armyagov/Shutterstock, p43bl ninikas/Shutterstock, p43br ninikas/Shutterstock, p44 stockfour/Shutterstock, p45tl Pavel Aleks/Shutterstock, p45tr oksana2010/Shutterstock, p44c 1Roman Makedonsky/Shutterstock, p44bcl M88/Shutterstock, p44br Africa Studio/Shutterstock, p44b Jovan Svorcan, p46bl Africa Studio/Shutterstock, p46br Attapol Yiemsiriwut/Shutterstock, p47 Jim Lipschutz/Shutterstock.com, p50 Brittny/Shutterstock, p51t Petr Malyshev/Shutterstock, p51c Greg Browning/Shutterstock, p51b Somchai Som/Shutterstock, p53 YiAN Kourt/Shutterstock, p54t MishAl/Shutterstock, p55 Alexey Stiop/Shutterstock, p56l Aleksander Bolbot/Shutterstock, p56r ktsdesign/Shutterstock, p57 noina/Shutterstock, p58t Deyan Georgiev/Shutterstock, p58b Zholobov Vadim/Shutterstock, p60tr connel/Shutterstock, p61 Vadim Sadovski/Shutterstock, p62l Chockdee Permploysiri/Shutterstock, p62cl Haoka/Shutterstock, p62cr Marina Biryukova/Shutterstock, p62r Benevolente82/Shutterstock, p63 Ensuper/Shutterstock, p66l TaiChesco/Shutterstock, p66c Mark Humphreys/Shutterstock, p66r KENG MERRY/Shutterstock, p66br andras_csontos/Shutterstock, p67t Alex James Bramwell/Shutterstock, p67b Ian Grainger/Shutterstock, p68l Scott Rothstein/Shutterstock, p68r Scott Rothstein/Shutterstock, p69tl winnond/Shutterstock, p69tr Mega Pixel/Shutterstock, p69c sjgh/Shutterstock, p69bl Christophe Testi/Shutterstock, p69br ecco/Shutterstock, p73tl SARIN KUNTHONG/Shutterstock, p73tr AleksandrN/Shutterstock, p75 valleyboi63/Shutterstock, p77 fboudrias/Shutterstock, p82t R R/Shutterstock, p82b IgorZh/Shutterstock, p86l Roman Babakin/Shutterstock, p86c Juriah Mosin/Shutterstock, p86r alvarojavier/Shutterstock, p87 IgorZh/shutterstock, p89 Atakan Yildiz/Shutterstock, p90 Picsfive/Shutterstock, p93 xtock/Shutterstock, p94 GaroManjikian/Shutterstock, p95 MarcelClemens/Shutterstock.

predator	An animal that kills and eats other animals.
prediction	A statement about a future event or outcome that can be scientifically tested.
prescription	Something a doctor gives to a patient who requires a controlled drug or medicine; a written instruction that authorises the purchase of the drug.
prey	The creatures that an animal hunts and eats.
producer	A thing that makes or produces something.
product	A substance formed in a chemical reaction.
protect	To protect someone or something is to prevent them from being harmed or damaged.
quarantine	To keep apart from other living things in case of infection or disease.
random	Not regular.
ray diagram	A diagram showing the direction of light rays.
reflect	To bounce something back without absorbing it, for example light or sound.
regular	Arranged in a definite pattern.
rehabilitation	To restore an animal to health after injury or illness.
relax	To relax a muscle is to allow it to become loose and less tight. It is the opposite of contract.
reversible	An event or something that can be changed back to its previous state.
rotate	To turn about an axis.
safe	Not in any danger.
scientific enquiry	Research and testing that is done when you want to find out about something.
sensor	A device that detects an energy change, for example, movement, sound, light.
series circuit	An electrical circuit where the components are arranged one after the other in a continuous circuit.
shadow	The dark shape made on a surface when an object prevents light from reaching it.
simplified	Made easier.
skeleton	Your skeleton is the framework of bones in your body.
Solar System	The Sun and all the planets, comets and asteroids that orbit round it.
solidify	To change from a liquid into a solid. See also *freezing*.
sphere	A 3D shape; a ball is a sphere.
stamina	The ability to carry on doing something for a long time, for example physical exercise.
stored energy	Also known as potential energy, the energy stored in a non-moving object.
stress	A force that produces a strain on something.
substance	Matter which has a specific composition and properties.
support	A thing that bears the weight of something or keeps it upright.
switch	An on–off control for an electrical device or machine.
tectonic plates	The Earth's crust is made up of large, moving pieces called plates. All of the Earth's land and water sit on these plates. The plates are made of solid rock.
transferred	Moved from one place to another.
tsunami	A giant destructive wave caused by an earthquake under the ocean.
Universe	Everything that exists.
vaccination	To be given a vaccine.
vaccine	A medicine or drug used to provide immunity from particular diseases.
vertebrates	Animals with a backbone (spine).
veterinary medicine	To do with diseases, injuries and treatments of animals.
vibrating	Moving a tiny amount backwards and forwards very quickly.
volcano	A mountain or hill on the Earth's crust where lava comes out.
wasted energy	Energy that is not usefully transferred.
k	Energy transfer that occurs when an object is moved.

galaxy	A system of billions of stars and planetary bodies held together by a gravitational pull.
gravity	The force that makes things fall when you drop them.
habitat	The natural home of a plant or animal.
heat	To make something warmer, or to have warmth or the quality of being hot.
herbivore	An animal that eats plants.
horizon	The line where the Earth's surface and the sky appear to meet.
insulator	An object or type of material that does not allow the flow of electric current.
invertebrates	Animals without a backbone.
investigate	To discover or examine the facts of something or a situation.
irreversible	Something that is not able to be reversed.
joints	The position in the body where two parts of the skeleton fit together, for example the knee or elbow joints.
lava	Hot molten rock that comes out of the top of a volcano.
light	The brightness that comes from things such as the Sun, fire or lamps that enables you to see things; a lamp or other device that gives out brightness.
light source	A device or natural feature that is a source of light.
lightning	A natural electrical discharge between the clouds and the ground.
mantle	The layers of the Earth between the crust and the core.
material	A substance from which something is made.
medicine	Medicine is the treatment of illness and injuries by doctors and nurses. Medicines are drugs used to treat illness.
melting	Something changing from a solid to a liquid.
model	To imitate or use something as an illustration of the real thing; two- or three-dimensional representations of actual living or non-living things.
molten rock	Rock that has become heated and turned to a liquid.
moon/Moon	A large natural object in orbit around a planet. The Moon is the name given to the moon that orbits the Earth.
movement	The action of moving.
movement energy	Energy that is stored in moving objects.
muscles	Bands of fibrous tissues in the body that contract to allow movement.
omnivore	An animal that eats both meat and plants.
orbit	The curved path followed by an object going around the Sun or a planet.
organs	The parts of an organism that have specific and vital functions, for example the heart or liver.
particle	An extremely small piece of matter.
particle model	A theory which helps us explain how matter moves, changes state and behaves.
pattern	A sequence that reoccurs.
pesticides	Poisons used on crops to kill pests.
pharmacy	A shop where you can buy medicines.
photosynthesis	The process by which plants make their food using sunlight, carbon dioxide and water.
physical changes	A change from one state to another without a chemical reaction taking place; a reversible change.
planet	A round object in space which orbits the Sun or another star and is lit by light from it.
planetary system	A group of planetary bodies in orbit around a star system.
plaster of Paris	Powdered gypsum (a type of rock) that sets hard when mixed with water.
plastic	Plastics are human-made materials, for example polythene and PVC.
potential energy	Also known as stored energy, the energy stored in a non-moving object.

Glossary

adapted	Something that has changed to suit its conditions or needs.
aftershock	A small earthquake than happens after a main earthquake.
angle	The amount of turn between two lines that meet at a point, usually measured in degrees.
asteroid	A small rocky object that orbits the Sun.
axis	An imaginary line around which something rotates.
backbone	A word used to describe the spine. All vertebrates have a backbone.
bounce	When something moves quickly up, back or away from a surface after hitting it.
break	A break in a circuit is an interruption in the flow of electricity; to break something is to shatter it into pieces.
carnivore	An animal that eats meat.
chemical reaction	A chemical change that results in a different substance being made.
circuit	An electrical circuit is a complete route around which an electric current can flow.
comet	A small object with an icy core that orbits the Sun; when a comet passes near to the Sun it has 'tail' of gases.
commercial	Something that is sold or made to sell for a profit.
component	A part or element of something, for example a part of a machine.
composition	What something is made from.
compound	A pure substance made from all the same molecules.
compress	To squash something to make it smaller.
condensation	Condensation is a coating of tiny drops of water formed on a surface by steam or vapour.
conductor	An object or type of material that allows the flow of electric current.
consumer	An animal or thing that eats or uses something.
contract	To become smaller. To contract a muscle is to allow it to become firm and tight, and smaller. It is the opposite of relax.
cooling	The process of becoming cool.
core	The Earth's central layer.
crust	The Earth's hard outer layer.
current	A flow of electricity.
disease	An illness that can be spread easily; caused by a germ.
drugs	A drug is any substance that affects how your body works, for example aspirin.
element	A pure substance in its simplest form that cannot be broken down any further.
energy	The ability to be active or of doing work.
energy saver	An efficient electrical device.
environment	The surroundings and conditions in which something lives.
evaporation	When something changes gradually from a liquid into a gas.
exercise	To do physical work to improve health and fitness.
exoskeleton	The hard outer shell of some animals.
fair	A fair test is when you change only the thing you are measuring and keep all other conditions the same.
fault line	Where the edges of the Earth's tectonic plates meet.
flexible	The property of materials such as rubber that you can bend easily; the ability to bend easily.
freezing	Changing from a liquid into a solid; extremely cold. See also *solidify*.